W9-BZX-047

McGraw-Hill
My Math

CCSS

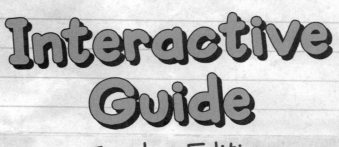

Interactive
Guide

Teacher Edition

Grade 1

Mc
Graw
Hill
Education

ConnectED.mcgraw-hill.com

STEM McGraw-Hill is committed to providing instructional materials in Science, Technology, Engineering, and Mathematics (STEM) that give all students a solid foundation, one that prepares them for college and careers in the 21st century.

Send all inquiries to:
McGraw-Hill Education
8787 Orion Place
Columbus, OH 43240

Selections from:
ISBN: 978-0-02-132705-8 *(Grade 1 Student Edition)*
MHID: 0-02-132705-X *(Grade 1 Student Edition)*
ISBN: 978-0-02-130895-8 *(Grade 1 Teacher Edition)*
MHID: 0-02-130895-0 *(Grade 1 Teacher Edition)*

Printed in the United States of America.

Visual Kinesthetic Vocabulary® is a registered trademark of Dinah-Might Adventures, LP.

2 3 4 5 6 7 8 9 ROV 20 19 18 17 16 15

Contents

Chapter 3 Addition Strategies to 20

Chapter 4 Subtraction Strategies to 20

Chapter 5 Place Value

Chapter 10 Three-Dimensional Shapes

Proficiency Level Descriptors

	Interpretive (Input)		Productive (Output)	
	Listening	**Reading**	**Writing**	**Speaking**
An Emerging Level EL • New to this country; may have memorized some everyday phrases like, "Where is the bathroom", "My name is….", may also be in the "silent stage" where they listen to the language but are not comfortable speaking aloud • Struggles to understand simple conversations • Can follow simple classroom directions when overtly demonstrated by the instructor	• Listens actively yet struggles to understand simple conversations • Possibly understands "chunks" of language; may not be able to produce language verbally	• Reads familiar patterned text • Can transfer Spanish decoding somewhat easily to make basic reading in English seem somewhat fluent; comprehension is weak	• Writes labels and word lists, copies patterned sentences or sentence frames, one- or two-word responses	• Responds non-verbally by pointing, nodding, gesturing, drawing • May respond with yes/no, short phrases, or simple memorized sentences • Struggles with non-transferable pronunciations.
An Expanding Level EL • Is dependent on prior knowledge, visual cues, topic familiarity, and pretaught math-related vocabulary • Solves word problems with significant support • May procedurally solve problems with a limited understanding of the math concept.	• Has ability to understand and distinguish simple details and concepts of familiar/ previous learned topics	• Recognizes obvious cognates • Pronounces most English words correctly, reading slowly and in short phrases • Still relies on visual cues and peer or teacher assistance	• Produces writing that consists of short, simple sentences loosely connected with limited use of cohesive devices • Uses undetailed descriptions with difficulty expressing abstract concepts	• Uses simple sentence structure and simple tenses • Prefers to speak in present tense.
A Bridging Level EL • May struggle with conditional structure of word problems • Participates in social conversations needing very little contextual support • Can mentor other ELs in collaborative activities.	• Usually understands longer, more elaborated directions, conversations, and discussions on familiar and some unfamiliar topics • May struggle with pronoun usage	• Reads with fluency, and is able to apply basic and higher-order comprehension skills when reading grade-appropriate text	• Is able to engage in writing assignments in content area instruction with scaffolded support • Has a grasp of basic verbs, tenses, grammar features, and sentence patterns	• Participates in most academic discussions on familiar topics, with some pauses to restate, repeat, or search for words and phrases to clarify meaning.

Strategies for EL Success

Surprisingly, content instruction is one of the most effective methods of acquiring fluency in a second language. When content is the learner's focus, the language used to perform the skill is not consciously considered. The learner is thinking about the situation, or how to solve the problem, not about the grammatical structure of their thoughts. Attempting skills in the target language forces the language into the subconscious mind, where useable language is stored. A dramatic increase in language integration occurs when multiple senses are involved, which causes heightened excitement, and a greater student investment in the situation's outcome. Given this, a few strategies to employ during EL instruction that can make teaching easier and learning more efficient are listed below:

- Activate EL prior knowledge and cultural perspective
- Use manipulatives, realia, and hands-on activities
- Identify cognates
- Build a Word Wall
- Modeled talk
- Choral responses
- Echo reading
- Provide sentence frames for students to use
- Create classroom anchor charts
- Utilize translation tools (i.e. Glossary, eGlossary, online translation tools)
- Anticipate common language problems

Common Problems for English Learners

Phonics Transfers	Cantonese	Haitian Creole	Hmong	Korean	Spanish	Vietnamese
Pronouncing the /k/ as in cake	●		●	●		
Pronouncing the digraph /sh/	●		●		●	●
Hearing and reproducing the /r/, as in *rope*	●		●	●		●
/j/			●		●	
Hearing or reproducing the short /u/		●	●			
Grammar Transfers						
Adjectives often follow nouns		●	●		●	●
Adjectives and adverb forms are interchangeable		●	●			
Nouns have feminine or masculine gender					●	
There is no article or there is no difference between articles *the* and *a*		●	●			●
Shows comparative and superlative forms with separate words			●		●	
There are no phrasal verbs				●	●	

How to Use the Teacher Edition

The Interactive Guide provides scaffolding strategies and tips to strengthen the quality of mathematics instruction. The suggested strategies, activities, and tips provide additional language and concept support to accelerate English learners' acquisition of academic English.

English Learner Instructional Strategy

Each lesson – including Problem Solving – references an English Learner Instructional Strategy that can be utilized before or during regular class instruction. These strategies specifically support the Teacher Edition and scaffold the lesson for English learners (ELs).

Categories of the scaffolded support are:
- Vocabulary Support
- Language Structure Support
- Sensory Support
- Graphic Support
- Collaborative Support

The goal of the scaffolding strategies is to make each individual lesson more comprehensible for ELs by providing visual, contextual and linguistic support to foster students' understanding of basic communication in an academic context.

Lesson 2 Model Addition
English Learner Instructional Strategy

Graphic Support: Graphic Organizers

Use a large presentation part-part-whole mat (or draw one) and pictures to demonstrate the lesson. Have a picture of 2 girls, a picture of 1 boy, and pictures of 3 tents ready for the Model Addition activity.

Say, *Two girls bought tents from a store.* Place the picture of the girls in the top left part of the mat and write the number 2. Say, *One boy bought a tent from the same store.* Place the picture of the boy in the top right part of the mat. Write, 1.

Say, *I want to know how many people bought tents in all. You can add the numbers from each part to find the whole.* Place the image of three tents in the bottom part of the mat and write the number 3.

Provide this sentence frame for students to report the number of parts in the whole: **The parts are _____ and _____. The whole is _____.**

Since ELs benefit from visual references to new vocabulary, many of the English Learner Instruction Strategies suggest putting vocabulary words on a Word Wall. Choose a location in your classroom for your Word Wall, and organize the words by chapter, by topic, by Common Core domain, or alphabetically.

How to Use the Teacher Edition *continued*

English Language Development Leveled Activities

These activities are tiered for Emerging, Expanding, and Bridging leveled ELs. Activity suggestions are specific to the content of the lesson. Some activities include instruction to support students with lesson-specific vocabulary they will need to understand the math content in English, while other activities teach the concept or skill using scaffolded approaches specific to ELs. The activities are intended for small group instruction, and can be directed by the instructor, an aide, or a peer mentor.

English Language Development Leveled Activities

Emerging Level	Expanding Level	Bridging Level
Exploring Language Structure	**Build Background Knowledge**	**Act It Out**
Write both the singular and plural form of the following words on the board: *coin/ coins, penny/pennies, cent/ cents* and *dime/dimes*. Demonstrate the value of pennies and dimes by showing that ten pennies is equal to one dime. Use money to model and say, *This is one* **coin**. *Here are two* **coins**. *One penny is one* **cent**. *Two* **pennies** *is two* **cents**. *One* **dime** *is ten* **cents**. *Two* **dimes** *is twenty* **cents**. Point to the singular or plural form of the listed words as you say each. Have students recite each.	Distribute 10 pennies and 1 dime to students. Say, *One penny* **equals** *one cent. Two pennies* **equals** *two cents. When amounts are equal, they are the same. Two* _____ ave _____ sentence and hold up one or two pennies. Say, *Ten pennies* **equals** *one dime. A dime* **is equal to** *ten pennies. Ten pennies is the same amount as one dime.* Have students count out 10 pennies and repeat each sentence as they point to the coins.	Display 10 pennies and 1 dime. Say, *Ten pennies equals one dime. Ten pennies is the same amount as one dime.* Use dimes to model numbers 10, 20, 40, 70, identifying each set using the sentence frames: _____ *dimes is equal to* _____ *tens* and so on. Distrib _____ manipulative dimes t _____ student. Say, *Show th* _____ *tens.* Then have them _____ using dimes and the sentence frame: **____ dimes is equal to ____ tens.** Repeat with different numbers.

> Teacher talk is *gray italicized*.

> Student talk is **boldfaced**.

Multicultural Teacher Tip

Because many word problems involve prices and/or determining changes in monetary value, ELs will benefit from an increased understanding of American coins and bills. A chart or other kind of graphic organizer visually comparing coin and bill values and modeling how to write dollars and cents in decimal form would help these students. You may also want to have ELs describe the monetary systems of their native countries. Identifying similarities or differences with the American system can help familiarize students with dollars and cents.

Multicultural Teacher Tip

These tips provide insight on academic and cultural differences you may encounter in your classroom. While math is the universal language, some ELs may have been shown different methods to find the answer based on their native country, while cultural customs may influence learning styles and behavior in the classroom.

What's the Math in this Chapter?

Mathematical Practice

The goals of the Mathematical Practice activity are to help students clarify the specific language of the Mathematical Practice and rewrite the Mathematical Practice in simplified language that students can **relate** to. Examples are discussed to make connections, showing students how they use this Mathematical Practice to solve math problems.

Chapter 10 Three-Dimensional Shapes
What's the Math in This Chapter?

Mathematical Practice 7: Look for and make use of structure

Create and then present a composite object made of at least one cube, cylinder, and cone. Ask, *What shapes do you see?* Allow students time to respond. Remind them of the definition of structure. Say, *The structure of this object includes a cube, cylinder, and cone.* Then dismantle the object, identifying each three-dimensional shape.

Ask, *Have you ever noticed that objects are made of shapes?* Solicit **Yes** from students. Display a three-dimensional shape and ask, *Can you find an object in our classroom that has this structure?* Allow students time to find real-world objects that match, then repeat with another three-dimensional shape.

Ask, *How can structure in math help us?* ____ peer, then as a group. The goal is to reco____ structure can help them identify three-dim____

Display the chart with Mathematical Prac____ Mathematical Practice 7 and have students assi__ in rewriting an additional "I can" statement, for example: **I can see the structure of shapes in real objects.** Have students find and cut out images of three-dimensional shapes and add them to the chart.

Inquiry of the Essential Question:

How can I identify three-dimensional shapes?

Inquiry Activity Target: **Students come to a conclusion that structure can be used to ____derstand the shape of real-world objects.**

____ chapter, present the Essential Question to ____phic organizer will offer opportunities for students ____es, and apply prior knowledge of three-____senting the Essential Question. As they investigate, ____aw, write, and collaborate with peers to demonstrate their observations and thinking. Then have students present additional questions they may have to a peer to extend discussions.

> Mathematical Practice is rewritten as an "I can" statement.

> Inquiry Activity Target connects Mathematical Practice to Essential Question.

Inquiry of the Essential Question

As an introduction to the Chapter, the Inquiry of the Essential Question graphic organizer activity is designed to introduce the Essential Question. The activity offers opportunities for students to observe, make inferences, and apply prior knowledge of samples/models representing the Essential Question. Collaborative conversations drive students toward the Inquiry Activity Target which is to make a connection between the "Mathematical Practice of the chapter" and the "Essential Question of the chapter."

How to Use the Student Edition

Each student page provides EL support for vocabulary, note taking, and writing skills. These pages can be used before, during, or after each classroom lesson. A corresponding page with answers is found in the Teacher Edition.

Inquiry of the Essential Question

Students observe, make inferences, and apply prior knowledge of chapter specific samples/models representing the Essential Question of the chapter. Encourage students to have collaborative conversations as they share their ideas and questions with peers. As students inquire the math models, present specific questions that will drive students toward the Inquiry Activity Target which is stated on the Teacher Edition page.

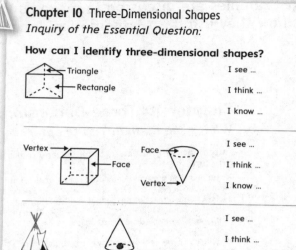

Chapter 10 Three-Dimensional Shapes
Inquiry of the Essential Question:

How can I identify three-dimensional shapes?

Triangle
Rectangle

I see …
I think …
I know …

Vertex
Face
Face
Vertex

I see …
I think …
I know …

Cone
Cone

I see …
I think …
I know …

Questions I have…

Cornell Notes/Note Taking

Cornell notes offer students a method to use to take notes, thereby helping them with language structure. Scaffolded sentence frames are provided for students to fill in important math vocabulary by identifying the correct word or phrase according to context. Encourage students to refer to their books to locate the words needed to complete the sentences. Each note taking graphic organizer will support students in answering the Building on the Essential Question.

Lesson 6 Note Taking
Read Bar Graphs

Read the question. Write words you need help with.
Use your lesson to write your Cornell notes.

Building on the Essential Question	Notes:
How can I read bar graphs?	I know the bars on a bar graph tell _____
	The _____ on a bar graph can be horizontal or vertical.
	I should look where each bar _____.
Words I need help with:	Then I should read the _____.

Favorite Fruit

horizontal bars

0 1 2 3 4 — numbers

Four-Square Vocabulary

Four-square vocabulary reinforces the lesson by having students write a definition utilizing the Glossary, write a sentence using the vocabulary in context, and create an example of the vocabulary. Suggest that students use translation tools and write notes in English or in their native language on the graphic organizer as well for clarification of terms. A blank Four-Square Vocabulary template has been provided on p. xix for use with other lessons.

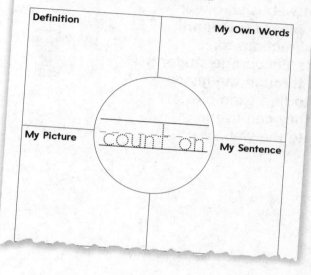

Lesson 2 Four-Square Vocabulary
Count On Using Pennies

Trace the word. Write the definition for *count on.* Write what the words mean, draw a picture, and write your own sentence using the words.

Definition	My Own Words
My Picture	My Sentence

count on

Vocabulary Sentence Frame

Sentence frames provide students written/oral language support giving students a framework to explain their thinking. Review the terms in the word back with students using models and math examples. Read all sentence frames aloud and encourage students to echo read. Have students review the lesson examples to assist them with sentence frame completion.

Lesson 10 Vocabulary Sentence Frames
Subtract from 8

The math words in the word bank are for the sentences below. Write the words that fit in each sentence on the blank lines.

Word Bank		
minus	difference	equals

I. The answer to a subtraction problem is the _____.
$8 - 5 = 3$

2. Eight minus four _____ four.

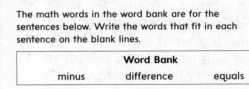

3. Eight _____ six equals two.

How to Use the Student Edition *continued*

Concept Web

Concept webs are designed to show relationships between concepts and to make connections. As each concept web is unique in design, please read and clarify directions for students. Encourage students to look through the lesson pages to find examples or words they can use to complete the web.

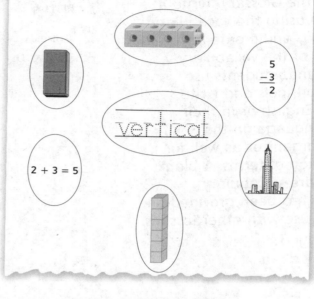

Lesson 5 Concept Web
Vertical Subtraction

Write *vertical* in the center oval. Draw lines to match the vertical items to the word *vertical*.

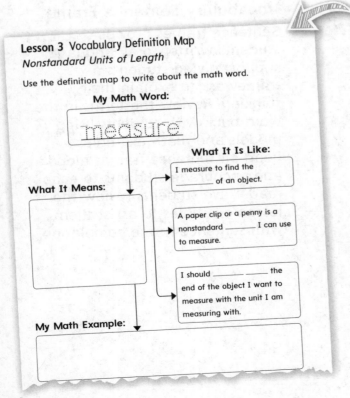

Lesson 3 Vocabulary Definition Map
Nonstandard Units of Length

Use the definition map to write about the math word.

My Math Word:
measure

What It Means:

What It Is Like:

I measure to find the _____ of an object.

A paper clip or a penny is a nonstandard _____ I can use to measure.

I should _____ the end of the object I want to measure with the unit I am measuring with.

My Math Example:

Definition Map

The definition maps are designed to address a single vocabulary word, phrase, or concept. Students should use the Glossary to help define the word in the description box. Sentence frames are provided to scaffold characteristics from the lesson. Students can refer to the lesson examples and Glossary to assist them in completing the sentence frames as well as creating their own math examples. Make sure you review with students the tasks required. A blank vocabulary definition map template can be found on page xviii for use with other lessons.

Problem Solving

Each Problem Solving page focuses on scaffolding Exercise 1 from the Homework portion of the lesson in the book. The text for the exercise highlights signal words and phrases to help students decipher the key information in the problem. Visual images as well as tables and part-part-whole mats are included to assist students with the problem-solving process. Sentence frames of the restated question are provided for student oral practice. A blank Problem Solving template can be found on page xx in this book if students need additional assistance with other exercises.

Lesson 4 Problem Solving

STRATEGY: Write a Number Sentence

<u>Underline</u> what you know. (Circle) what you need to find. Write a subtraction number sentence to solve.

1. There are **9** flamingos in the water.

   5 flamingos **get out.**

   How many flamingos **are still in the water?**

   flamingo

   water

Part	Part
Whole	

   ___ ⊖ ___ ⊜ ___

___ flamingos **are still in the water.**

English/Spanish Used in Grade 1

Chapter	English	Spanish
1	add	sumar
	addition number sentence	enunciado de suma
	equal (=)	igual (=)
	false	falso
	in all	en total
	part	parte
	plus (+)	más (+)
	same	el mismo
	sum	suma
	true	verdadero
	whole	total
	zero	cero
2	compare	comparer
	difference	differencia
	minus (−)	menos (−)
	subtraction number sentence	enunciado de resta
	related facts	operaciones relacionadas
	subtract	restar
	subtraction	sustracción
3	addend	sumando
	count on	seguir contando
	doubles	dobles
	doubles minus 1	dobles menos 1
	doubles plus 1	dobles más 1
	number line	recta numérica
4	count back	contar hacia atás
	fact family	familia de operaciones
	missing addend	sumado desconocido
5	equal to (=)	igual a (=)
	greater than (>)	mayor que (>)
	hundred	centena
	less than (<)	menor que (<)
	ones	unidades
	regroup	reagrupar
	tens	decenas
6	*all vocabulary are review words*	
7	bar graph	gráfica de barras
	data	datos
	graph	gráfica

Chapter	English	Spanish
	picture graph	gráfica con imágenes
	survey	encuesta
	tally chart	table de conteo
8	analog clock	reloj analógico
	digital clock	reloj digital
	half hour	media hora
	hour	hora
	hour hand	manecilla horaria
	length	longitud
	long	largo
	measure	medir
	minute	minuto
	minute hand	minutero
	o'clock	en punto
	short	corto
	unit	unidad
9	circle	círculo
	composite shape	figura compuesta
	equal part	partes iguales
	fourths	cuartos
	halves	mitades
	rectangle	rectángulo
	side	lado
	square	cuadrado
	trapezoid	trapecio
	triangle	triángulo
	two-dimensional shapes	figura bidimensional
	vertex	vértice
	whole	el todo (entero)
10	cube	cubo
	cone	cono
	cylinder	cilindro
	face	cara
	rectangular prism	prisma rectangular
	three-dimensional shape	figura tridimensional

Vocabulary Definition Map

Use the definition map to write a description and list characteristics about the vocabulary word or phrase. Write or draw a math example.

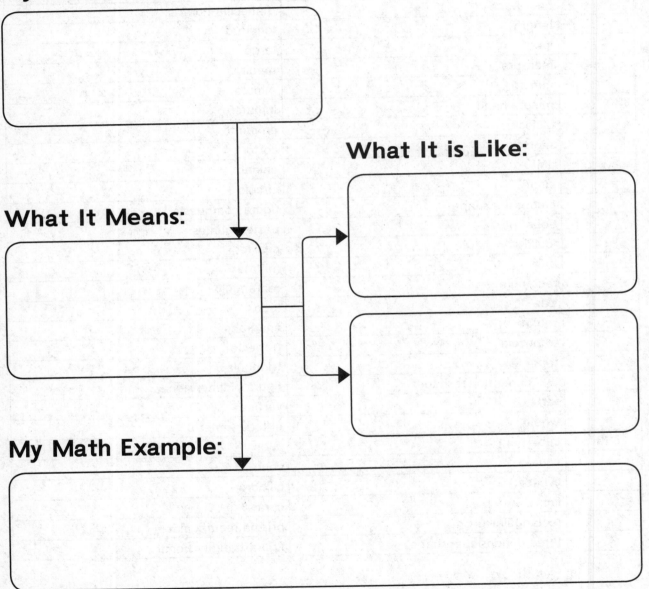

My Math Word:

What It is Like:

What It Means:

My Math Example:

Teacher Directions: Provide a description, explanation, or example of the new term using images or real objects. Have students use the lesson or Glossary to define the math term. Direct students to list characteristics, and draw a picture representing their math term. Then encourage students to describe their picture to a peer.

Four-Square Vocabulary

Write the definition for the math word. Write what the word means in your own words. Draw a picture that shows the math word meaning. Then write your own sentence using the word.

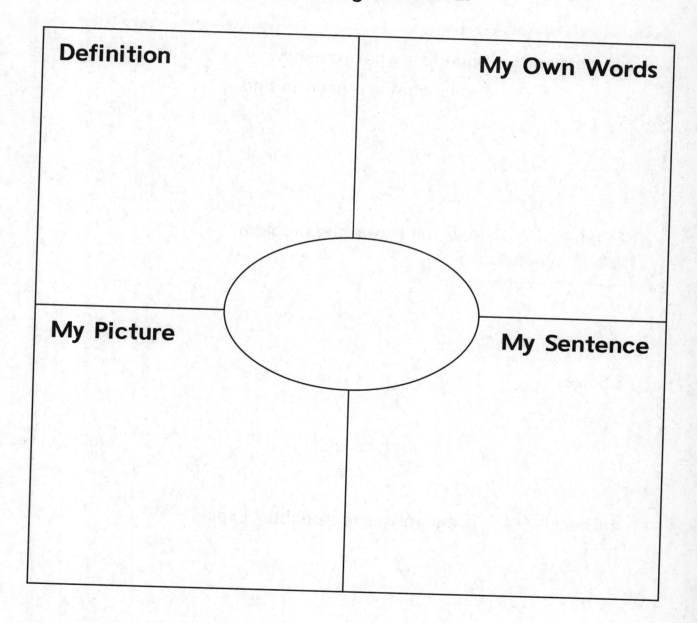

Definition

My Own Words

My Picture

My Sentence

Teacher Directions: Provide a description, explanation, or example of the new term using images or real objects. Have students use the Glossary to write the definition. Direct students to write a definition in their own words and draw a picture representing their math term. Have students write a sentence using the term and then encourage students to read their sentence to a peer.

Problem Solving

_____

_____

_____

_____

1 Understand   Underline what you know.

Circle what you need to find.

2 Plan   How will I solve the problem?

3 Solve   I will...

4 Check   Is my answer reasonable? Explain.

Chapter 1 Addition Concepts

What's the Math in This Chapter?

Mathematical Practice 6: Attend to Precision

Ask a student volunteer to draw a circle on the board. Then ask the student, *Why did you draw a circle on the board?* The student's response should be, **Because you told me to**. Then say, I wanted you to draw a square. Elicit a response from student such as, **You said to draw a circle.** Then say, *I'm sorry. I should have been more precise. I need to be careful when I talk about math. I meant to say square not circle.* Then ask the student to please erase the circle and draw a square. Say, *When I am not precise or careful, I can make mistakes.*

Ask, *Have you used math words like square and circle to explain something to a friend?* Solicit **Yes** from students, then have students share the various math vocabulary they have used thus far such as: *count, more, less, plus, minus, join, take away,* etc.

Ask, *Why should we be careful when we talk about math?* Encourage students to turn and talk with a peer. Then discuss as a group. The discussion goal is to have students recognize if they are not "precise/clear/careful" then mistakes can be made.

Display a chart with Mathematical Practice 6. Restate Mathematical Practice 6 and have students assist in rewriting it as an "I can" statement, for example: **I can be careful when I talk about math.** Have students draw or write examples of using math vocabulary precisely/clearly. Post the chart in the classroom.

Inquiry of the Essential Question:

How do I add numbers?

Inquiry Activity Target: **Students come to a conclusion that precision is necessary to join parts to make a whole.**

As an introduction to the chapter, present the Essential Question to students. The inquiry graphic organizer will offer opportunities for students to observe, make inferences, and apply prior knowledge of joining parts to make a whole representing the Essential Question. As they investigate, encourage students to draw, write, and collaborate with peers to demonstrate their observations and thinking. Then have students present additional questions they may have to a peer to extend discussions.

Regroup students and restate Mathematical Practice 6 and the Essential Question. Pose questions to reflect on what has been learned to guide students in making connections between the Mathematical Practice and the Essential Question.

NAME _____ DATE _____

Chapter 1 Addition Concepts
Inquiry of the Essential Question:

How do I add numbers?

● Part	Part
●●●●	●
Whole	
●●●● ●	

I see ...

I think ...

I know ...

Part	Part
3	4
Whole	
7	

I see ...

I think ...

I know ...

I see ...

●●● + ●● = ●●●●●

I think ...

I know ...

Questions I have...

_____

- - - - - - - - - - - - - - - - - - -

_____

_____

- - - - - - - - - - - - - - - - - - -

_____

Teacher Directions: Read the Essential Question for students. Have students echo read. Direct students to describe their observations, inferences, and prior knowledge of each math example. Encourage students to write or draw additional questions they may have. Then have students share their thinking/questions with a peer.

Grade 1 · **Chapter 1** *Addition Concepts* **1**

Lesson 1 Addition Stories

English Learner Instructional Strategy

Language Structure Support: Tiered Questions

Have a group of students come to the front of the classroom. Be sure the total number does not exceed 9 and includes boys and girls. Distribute sticky notes to model (Spanish cognate *modelo*) an addition story. Give the girls one color and the boys a different color. Have students put the sticky notes on the board, creating two separate groups. Discuss using tiered questions.

For emerging students, ask questions that can be answered with a gesture or one word such as, *How many "girl" sticky notes are there? and How many are there in all?*

For expanding/bridging students, ask questions that encourage a simple or complex sentence response such as, *How do you know this is a part? and Can you tell me a number story about the sticky notes?*

English Language Development Leveled Activities

Emerging Level	Expanding Level	Bridging Level
Word Recognition Write *number* and its Spanish cognate, *número* on a classroom cognate chart. Write the numbers 1–9 on the board. Count and point to each number. Circle the numbers and say, *These are numbers.* Write the alphabet on the board. Have students sing the alphabet song along with you. Say, *These are letters.* Shake your head and say, *They are not numbers.* Emphasize *not.* Count familiar objects, such as number of fingers on one hand. Prompt students to repeat the numbers.	**Show What You Know** Display a picture showing two groups of the same objects. Be sure the total number of objects does not exceed 9. Point to the first group. Ask, *How many in this group?* Repeat for the second group. Distribute counters and say, *Use counters to show how many you see in each group.* Be sure students group their counters into two separate groups. Have students describe the total number of counters using the sentence frame: **I have ____ counters in all.**	**Making Connections** Group students into multilingual pairs. Distribute counters and say, *Use the counters to show an addition story.* Be sure students group their counters in two separate groups. Explain to students that they will tell a story about each group of counters. The stories may be make-believe or based on their own personal experiences. Ask each pair of students to explain their addition story to another pair or small group of students.

Teacher Notes:

NAME _____ DATE _____

Lesson 1 Four-Square Vocabulary
Addition Stories

Trace the word. Write the definition for *number*.
Write what the word means, draw a picture, and
write your own sentence using the word.

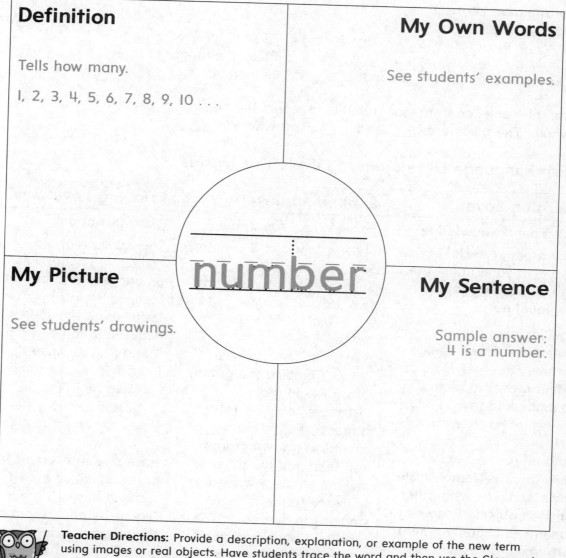

Definition

Tells how many.

1, 2, 3, 4, 5, 6, 7, 8, 9, 10 . . .

My Own Words

See students' examples.

number

My Picture

See students' drawings.

My Sentence

Sample answer:
4 is a number.

 Teacher Directions: Provide a description, explanation, or example of the new term using images or real objects. Have students trace the word and then use the Glossary to write the definition. Direct students to write a definition in their own words and draw a picture representing their math term. Have students write a sentence using the term. Then encourage students to read their sentence to a peer.

2 **Grade 1 · Chapter 1** *Addition Concepts*

Lesson 2 Model Addition
English Learner Instructional Strategy

Graphic Support: Graphic Organizers

Use a large presentation part-part-whole mat (or draw one) and pictures to demonstrate the lesson. Have a picture of 2 girls, a picture of 1 boy, and pictures of 3 tents ready for the Model Addition activity.

Say, *Two girls bought tents from a store.* Place the picture of the girls in the top left part of the mat and write the number 2. Say, *One boy bought a tent from the same store.* Place the picture of the boy in the top right part of the mat. Write, 1.

Say, *I want to know how many people bought tents in all. You can add the numbers from each part to find the whole.* Place the image of three tents in the bottom part of the mat and write the number 3.

Provide this sentence frame for students to report the number of parts in the whole: **The parts are _____ and _____. The whole is _____.**

English Language Development Leveled Activities

Emerging Level	Expanding Level	Bridging Level
Background Knowledge	**Show What You Know**	**Math in Action**
Show a paper circle. Say, *Whole,* as you outline the edge. Cut the circle into four equal parts.	Show a group of 8 counters. Say, *This is a whole.* Separate the counters into two groups. Say, *Each of these groups is part of the whole. When I add the parts together, it makes a whole.* Combine the groups into a whole group. Distribute counters. Have students separate the counters into two groups and then add the parts to make a whole. Provide this sentence frame to assist students in describing the process: **I have two parts. When I add the parts together, it makes a whole.**	Show a train of ten connecting cubes. Place it on one side of a bucket balance. Next, show two trains using ten connecting cubes in total (for example, 4 and 6) and place them on the other side of the bucket balance. Say, *This whole is the same as these two parts. When I add the parts, it makes a whole.* Have students repeat the bucket balance activity, working in small groups. Direct them to describe their models using the words: *whole, part,* and *add.*
Show 3 pieces and say, *Three parts.* Show 1 piece and say, *One part.* Hold up all four pieces to recreate the circle and say, *When I add the parts together I make a whole. Three parts plus one part makes a whole of four.* Demonstrate whole and part with other common objects. Encourage students to chorally repeat part-whole phrases.		

Teacher Notes:

NAME _____ DATE _____

Lesson 2 Note Taking
Model Addition

Read the question. Write words you need help with. Use your lesson and the word bank to write your Cornell notes. Write or draw math examples to explain your thinking.

Building on the Essential Question	**Notes:**
How can I model addition?	**Word Bank** objects part add whole
	A ___part___ is a set of objects I want to join.
	When I add, I want to find the ___whole___.
Words I need help with: See students' words.	The whole is all of the ___objects___ in a group.
	I must ___add___ the parts to find the whole.

My Math Examples:
See students' examples.

Part	Part
Whole	

 Teacher Directions: Read the Building on the Essential Question and have students list words/phrases they need assistance with. Provide descriptions, explanations, or examples of the terms using images or real objects. Read each sentence frame and have students fill in the appropriate terms from the word bank. Have students read their notes aloud. Direct students to draw a picture representing the question. Then encourage students to describe their picture to a peer.

Grade 1 • Chapter 1 *Addition Concepts*   3

Lesson 3 Addition Number Sentences
English Learner Instructional Strategy

Language Structure Support: Multiple Meaning Word

Before the lesson, write the word sentence. Ask, *What is a sentence?* Accept and discuss all answers. Give examples and non-examples of a sentence, such as "I ate a bagel," as an example and "I drank a glass of," for a non-example. Briefly discuss why the non-examples are not sentences.

Write the phrase addition number sentence and next to it write $4 + 2 = 6$. Say, *This is an addition number sentence.* Ask, *What is this?* Have students respond chorally, **an addition number sentence.**

Write $3 + 2$. Ask, Is this an addition number sentence? **no** Ask, *What is it missing?* Sample answer: **It is missing an equals symbol and the sum.**

English Language Development Leveled Activities

Emerging Level	Expanding Level	Bridging Level
Number Knowledge Write *sum* and its Spanish cognate, *suma* on a cognate classroom chart. Write $3 + 2 = 5$ on the board. Represent the number sentence using groups of objects. Create a number sentence by counting each set, putting them together, and saying, *Three plus two equals five. The sum is five.* Have students repeat chorally. Distribute 9 connecting cubes to pairs and have them make two groups from the whole. Ask students to count each group, and identify the number sentence using the sentence frame: ____ **plus** ____ **equals** ____ .	**Word Recognition** Model an addition number sentence by writing $4 + 2 = 6$ and use counters to model it. Say, *The numbers and symbols represent the words, four plus two equals six. Six is the sum, or answer.* Distribute counters to pairs. Have each partner use counters to create a number sentence. Have each pair say their number sentence. Ask each pair, *What is the sum?* Then have students write the symbols for each part of the addition sentence and identify the plus and equals signs.	**Exploring Language Structure** Write the word *some* on the board. Pick up a handful of connecting cubes and say, *Here are some connecting cubes.* Emphasize *some*. Repeat with different items such as sheets of paper or manipulatives. Write $3 + 2 = 5$ and *sum* on the board. Say, *Three plus two equals five. The sum, or answer, to the addition problem is five.* Discuss that *some* and *sum* are called homophones. Have students write sentences using *some* and *sum*. Have them share their sentences with a peer.

Teacher Notes:

NAME _____ DATE _____

Lesson 3 Word Identification
Addition Number Sentences

Match each term to a symbol or number.

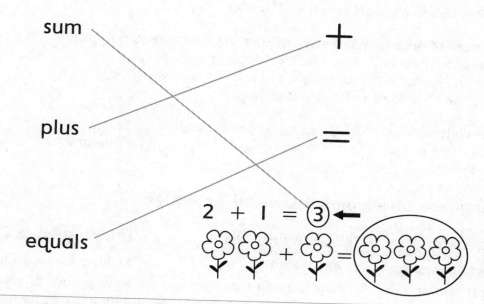

Write the correct term from above for each sentence on the blank lines.

addition number sentence
2 + 1 = 3

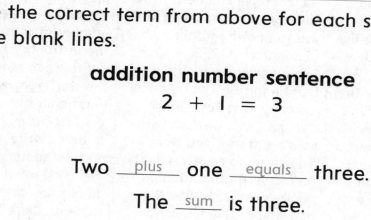

Two __plus__ one __equals__ three.

The __sum__ is three.

 Teacher Directions: Review the terms using manipulatives, such as counters or pennies. Have students say each word and then draw a line to match the word to its symbol or number. Direct students say the addition number sentence and then write the corresponding terms in the sentences. Encourage students to read the sentences to a peer.

Lesson 4 Add 0
English Learner Instructional Strategy

Sensory Support: Pictures and Photographs

Distribute write-on/wipe-off boards to each student. As an introduction to Explore and Explain, hold up a picture of seven people. Say, *Count the number of people in this picture. How many do you see?* **7** *Write the number to describe how many people are in the picture.* Have students write 7 on a write-on/ wipe-off board and display.

Hold up a picture of an empty table. Make sure the image does not have any people in it. Say, *Count the number of people in this picture. How many do you see?* **none** *Write a number to describe how many people are in the picture.* Have students write 0 and display.

Hold up both pictures. Ask, *Which picture has zero people in it? Write a number to describe the total number of people in both pictures.* Students write 7 and display.

English Language Development Leveled Activities

Emerging Level	Expanding Level	Bridging Level
Building Oral Language	**Act It Out**	**Making Connections**
Write *0* and *zero* and its Spanish cognate, *cero* on a cognate chart. Model the number zero by showing five counters and counting down as you take them away: *5, 4, 3, 2, 1.* As you remove the last one say, *Zero! How many total counters? Zero counters.* Sing and act out a counting down repetitive song, such as "Five Little Monkeys Jumping on the Bed." When there are zero monkeys left, stop and ask, *How many are left?* Have students answer, **Zero!**	Model adding zero by showing five counters. Have students chorally count with you to find the total number of counters: **1, 2, 3, 4, 5.** Show students an empty hand and say, *I'm going to add zero counters to the five counters.* Gesture with your hand as if you were adding more counters to the group. Write the addition number sentence 5 + 0 = and ask, *Now how many counters?* Have students chorally count with you to find the total number of counters.	Have groups of students play a board game in which the player moves a certain number of spaces. Explain the instructions for the game. Then model counting the starting space as "zero" and begin the counting on the first move. Have students write an addition number sentence describing their first move. When games are completed, have students share their game experience with the whole group.

Teacher Notes:

NAME _____ DATE _____

Lesson 4 Word Web
Add 0

Use the word web to show examples of zero.

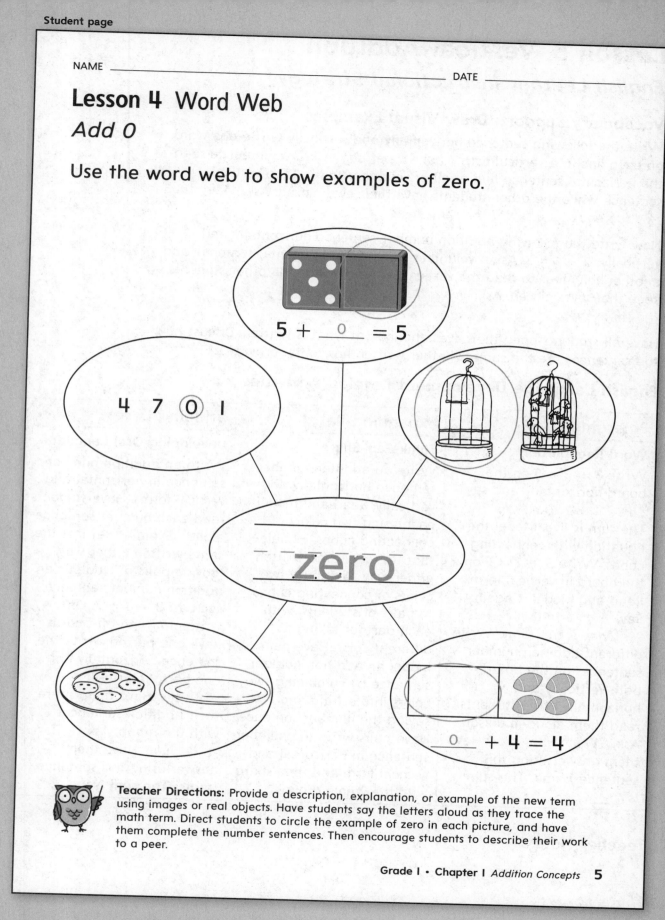

$5 + \underline{0} = 5$

4 7 0 1

zero

$\underline{0} + 4 = 4$

Teacher Directions: Provide a description, explanation, or example of the new term using images or real objects. Have students say the letters aloud as they trace the math term. Direct students to circle the example of zero in each picture, and have them complete the number sentences. Then encourage students to describe their work to a peer.

Grade 1 • Chapter 1 *Addition Concepts* **5**

Lesson 5 Vertical Addition

English Learner Instructional Strategy

Vocabulary Support: Draw Visual Examples

Write the following sentence horizontally and vertically (write one word on each line in a vertical list), *I have a pet dog.* Ask a volunteer to read the horizontal sentence and another volunteer to read the vertical sentence while the other students have their eyes closed. Ask, *Were the two sentences you heard the same?* **yes**

Now write the following addition number sentence horizontally and vertically, 2 + 1 = 3. Ask a volunteer to read the horizontal sentence and another volunteer to read the vertical sentence while the other students keep their eyes closed. Ask, *Were the two number sentences you heard the same?* **yes**

Have all students open their eyes and look at what is written. Discuss how the sentences are the same whether written horizontal or vertical.

English Language Development Leveled Activities

Emerging Level	Expanding Level	Bridging Level
Word Recognition Write *3 + 2 = 5* on the board horizontally and say, *Three plus two equals five.* The sum is five. Model the equation with connecting cubes. Write *3 + 2 = 5* on the board in vertical form. Read and model it again. Say, *Each addition number sentence is the same.* Write different addition number sentences on the board, both vertically and horizontally. Have students read them aloud in unison. Ask, *What is the sum?* Have them answer using this sentence frame, **The sum is ____.**	**Think-Pair-Share** Write an addition number sentence horizontally and vertically. Read and model each using two colors of connecting cubes. Explain that it is the same written either way. Distribute two colors of connecting cubes to pairs of students, with each partner taking a different color. Have pairs model an addition number sentence by combining two colors into a train and saying the number sentence. Have pairs write the number sentence in horizontal and vertical form and then share with the whole group.	**Developing Oral Language** Write an addition number sentence in horizontal and vertical forms. Have students read each number sentence aloud. Explain that it is the same written either way. Provide pairs of students an addition number sentence, such as: 3 + 4 = 7. On write-on/ wipe-off boards, have one partner write the sentence horizontally and one partner write it vertically. Have them share their addition sentences with the group. Ask, *What is the sum?* Have them answer using this sentence frame, **The sum is ____.**

Teacher Notes:

NAME _____ DATE _____

Lesson 5 Note Taking
Vertical Addition

Read the question. Write words you need help with. Use your lesson to write your Cornell notes. Write or draw math examples to explain your thinking. Share your examples with a classmate.

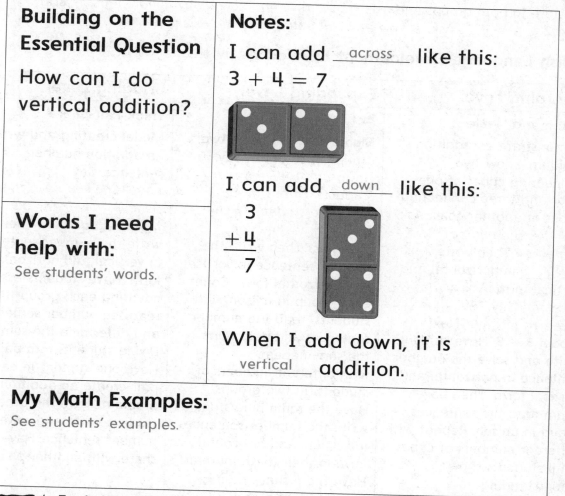

Building on the Essential Question	**Notes:**
How can I do vertical addition?	I can add ___across___ like this: 3 + 4 = 7
	I can add ___down___ like this: 3 + 4 —— 7
Words I need help with: See students' words.	When I add down, it is ___vertical___ addition.
My Math Examples: See students' examples.	

Teacher Directions: Read the Building on the Essential Question and have students list words/phrases they need assistance with. Provide descriptions, explanations, or examples of the terms using drawings or real objects. Read each sentence frame and have students write the appropriate terms. Have students read their notes aloud. Direct students to draw a picture representing the question. Then encourage students to describe their picture to a peer.

Lesson 6 Problem Solving
Strategy: Write a Number Sentence

English Learner Instructional Strategy

Collaborative Support: Act It Out

Select six students to act out the Problem Solving activity. Say, *There are two children fishing.* Have two of the students stand in a group and have them act out fishing. Say, *Four more children join them.* Have the remaining four students stand in a group and act out fishing. Ask, *How can we find how many children are fishing in all?* Gesture to the group of two and ask, *How many children are in this group?* **2** Gesture to the group of four and ask, *How many children are in this group, that joined them?* **4** Gesture to both groups and ask, *How do we find how many children in all?* **add** Write 2 + 4 = _____ Ask, *How many children are fishing in all?* **6**

English Language Development Leveled Activities

Emerging Level	Expanding Level	Bridging Level
Listen and Write	**Act It Out**	**Think-Pair-Share**
Demonstrate an addition problem. Show two connecting cubes of one color and three connecting cubes of another color. Ask, *What is the number sentence?* Then write 2 + 3 = _____ in horizontal and vertical form. Ask, *What is the sum?* Say, *Two plus three equals five.* Next model 3 + 5. Have students write and solve the number sentence in horizontal and vertical form. Then have them read the sentences aloud in unison. Repeat with different numbers of cubes until students show understanding.	Model addition using two groups of students. Say, *In this group I have three students. In this group I have two students. How many do we have all together?* Then write the number sentence in vertical and horizontal form. Count each group in unison with students, read the number sentence, and determine the sum. Repeat with different numbers of students in two groups. Have the remaining students write the number sentences in vertical and horizontal form in their math journals. Have them share their journals with a peer.	Model creating and writing an addition number sentence. Say, *I have two pencils. If you have three pencils, how many do we have all together?* Then write the number sentence in vertical and horizontal form. Have students assist counting each group, then read the number sentence and determine the sum. Divide students into pairs. Have one student in each pair create an addition problem. Have the other student write and solve the number sentence. Have them share with another pair.

Teacher Notes:

NAME _____ DATE _____

Lesson 6 Problem Solving
STRATEGY: Write a Number Sentence

Underline what you know. (Circle) what you need to find. Write an addition number sentence to solve.

I. **Andrew** saw **3** rabbits.

Tia saw **6** other rabbits.

How many rabbits did **they** (Andrew and Tia) see **in all?**

rabbit

Part	Part
Whole	

___3___ ⊕ ___6___ ⊜ ___9___

They saw ___9___ rabbits **in all**.

Teacher Directions: Provide a description, explanation, or example of the bold face terms and nouns using images or real objects. Read each sentence and have students echo read. Encourage students to use the part-part-whole mat, complete the addition sentence, and then write their answer in the restated question. Have students read the answer sentence aloud.

Grade 1 • **Chapter 1** Addition Concepts **7**

Lesson 7 Ways to Make 4 and 5

English Learner Instructional Strategy

Language Structure Support: Multiple-Meaning Word

Write the word *some* on the board. Write and then read aloud a sentence using the word some, such as, *I had some crackers with my soup.* Write the word *sum* and its Spanish cognate, *suma*. Write and then read aloud a sentence using the word sum, such as, *Two plus one equals three. The sum is three.*

Gesture to the words *some* and *sum* after each of the following questions. Ask, *Do these two words sound the same when read aloud?* **yes** *Do these two words have the same meaning?* **no** *Are these two words spelled the same?* **no** Define them as homophones.

Read aloud more example sentences using the word *some* or *sum*. Have students indicate which word was used in the sentence. Have bilingual pairs of students create sentences for each word: *some* and *sum*. Have pairs share their sentences with the class.

English Language Development Leveled Activities

Emerging Level	Expanding Level	Bridging Level
Act It Out	**Think-Pair-Share**	**Show What You Know**
Review the term *sum* and its Spanish cognate, *suma*. Select four students. Position them to show a combination of four with one in one group and three in another. Say, *One plus three equals four.* Have students repeat chorally. Say, *The sum is four.* Have students repeat. Continue positioning students in different combinations of four including $0 + 4 = 4$. After each number sentence is said aloud, name the sum. Repeat with combinations of five.	Model different ways to make four by showing combinations of counters. Use counters to demonstrate: $1 + 3 = 4$, $0 + 4 = 4$, and so on. After each model, write the number sentence on the board and ask, *What is the sum?* **4** Do the same for five. With each combination write the number sentence and ask, *What is the sum?* **5** Distribute counters and write-on/wipe-off boards. Have pairs show, write, and share the number sentences for combinations of four and five.	Have students work in groups of four or five and position themselves into combinations of four or five. Ask groups to record their combinations. Students should record the addition number sentence and sum using the sentence frames: ____ + ____ = ____. **The sum is** ____. As a class, discuss how to position classmates to reflect the different combinations of four or five, including $4 + 0 = 4$ and $5 + 0 = 5$. Discuss how the sums in a group are the same.

Teacher Notes:

NAME _____ DATE _____

Lesson 7 Sum Identification
Ways to Make 4 and 5

Circle all the ways to make 5 in the boxes below.

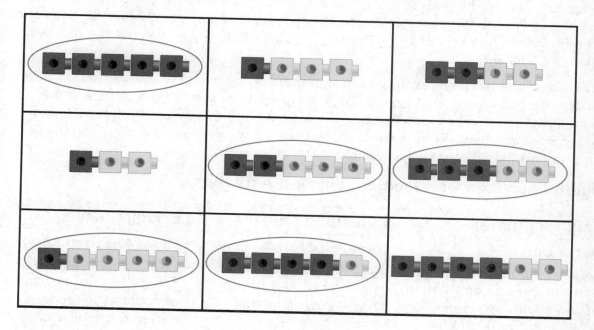

Draw a picture to show one way to make 4.

See students' examples.

Teacher Directions: Use manipulatives to show various ways to make 4 and 5. Model an addition sentence for each and have students repeat it. Have students circle each example that shows a sum of 5. Then direct students to draw a picture representing a way to make 4. Finally encourage students to describe their picture to a peer.

Lesson 8 Ways to Make 6 and 7

English Learner Instructional Strategy

Graphic Support: Graphic Organizers

Before the Explore and Explain lesson, display a large part-part-whole mat. Create pictures of 6 girls. Say, *We can use the part-part-whole mat to find the ways to make six. Imagine that six girl scouts are sharing two tents.* Place all 6 pictures of the girls in the top left part of the mat. Say, *Six girls are in the first tent and zero girls are in the second tent.* Provide this sentence frame for students to report the number for each part and the whole: **The parts are ____ and ____. The whole is ____.** Write the number sentence 6 + 0 = 6.

Move one of the girl pictures to the top right part of the mat. Say, *Five girls are in the first tent and one girl is in the second tent.* Have students identify the parts and the whole. Write the number sentence 5 + 1 = 6. Continue this way until you have found and recorded all the ways to make 6. Have small groups use manipulatives and Work Mat 3 to find the different ways to make 7.

English Language Development Leveled Activities

Emerging Level	Expanding Level	Bridging Level
Act It Out	**Think-Pair-Share**	**Show What You Know**
Select 7 students. Position them to show a combination of seven with two in one group and five in another. Say, *We will add 2 and 5.* Have students repeat the number sentence after you say it: **Two plus five equals seven.** Ask, *What is the sum?* 7 Continue positioning students to show different combinations of six and seven, including 6 + 0 = 6 and 0 + 7 = 7. After each number sentence, confirm that the sum is the same.	Use two colors of connecting cubes to demonstrate different combinations of six: 1 + 5 = 6, 0 + 6 = 6, and so on. After each model, say, *Now we will add the numbers.* Write each number sentence on the board, read it aloud, and ask, *What is the sum?* **The sum is six.** Distribute connecting cubes to pairs of students. Have pairs show and then write number sentences for different combinations of seven in their math journals. Have pairs share their combinations with the whole group.	Have students work in groups and use two colors of counters to model different combinations of six or seven. Students should record the addition number sentence and sum using the sentence frames: ____ + ____ = ____. **The sum is ____.** As a class discuss how to use the counters to reflect the different combinations of six or seven, including 6 + 0 = 6 and 7 + 0 = 7. Discuss how the sums of different combinations of six (and seven) are the same.

Teacher Notes:

NAME _____ DATE _____

Lesson 8 Sum Identification
Ways to Make 6 and 7

Circle all the ways to make 6 in the boxes below.

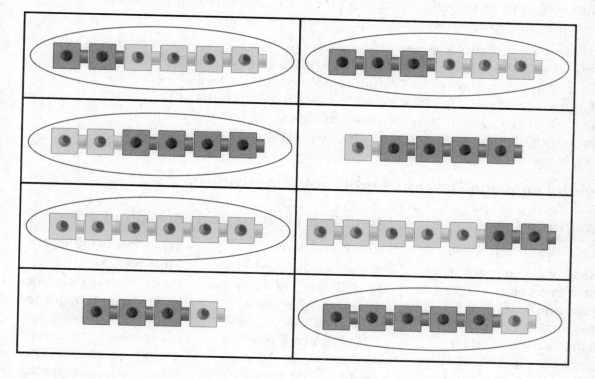

Draw a picture to show one way to make 7.

See students' examples.

 Teacher Directions: Use manipulatives to show various ways to make 6 and 7. Model an addition sentence for each and have students repeat it. Have students circle each example that shows a sum of 6. Then direct students to draw a picture representing a way to make 7. Finally encourage students to describe their picture to a peer.

Grade 1 • Chapter 1 *Addition Concepts* **9**

Lesson 9 Ways to Make 8

English Learner Instructional Strategy

Collaborative Support: Act It Out

Select 8 students to act out the Explore and Explain activity. Say, *There are five children swimming.* Have five of the students stand in a group and have them act out swimming. Say, *Three more children joined them.* Have the remaining three students stand in a group and act out swimming. Ask, *How can we determine how many children are swimming in all?*

Gesture to the group of five and ask, *How many children were swimming before more joined them?* **5** Gesture to the group of three and ask, *How many children joined them?* **3** Gesture to both groups and ask, *How many children are swimming in all?* **8** Write 5 + 3 = 8. Rearrange the 8 students and change the story to model the other ways to make eight. For example, no children are swimming and then eight children got in the water to swim will represent 0 + 8 = 8.

English Language Development Leveled Activities

Emerging Level	Expanding Level	Bridging Level
Show What You Know Model ways to make eight using two colors of counters. Show and write 3 + 5 = 8. Say, *Three plus five equals eight. The sum is eight. There are eight in all.* Have students repeat chorally. Write 4 + 4 = 8. Say, *Four plus four equals eight.* Have students repeat. Ask, *What is the sum?* **8** Gesture to each addition sentence and show that the sum for both number sentences are the same. Encourage students to say, **The sums are the same.** Continue this activity with other combinations of eight.	**Think-Pair-Share** Use connecting cubes to demonstrate: 1 + 7 = 8 and 2 + 6 = 8. After each model, ask, *How many in all?* **There are 8 in all.** *What is the sum?* **The sum is eight.** *Is the sum the same?* **yes** Distribute connecting cubes and write-on/wipe-off boards to pairs. Have pairs model and then write the number sentences for other combinations of eight. Have them share with the whole group using the sentence frames: ____ plus ____ equals ____. There are ____ in all. The sums are the ____.	**Exploring Language Structure** Write *eight* on the board. Have students count to eight and display eight using their fingers. Write *ate* on the board. Say, *I ate my breakfast.* Demonstrate *ate* by pantomiming eating food. Have students say, **eight** and **ate**, as you point to each word. Ask, *Do these words sound the same?* **yes** *Are these words spelled the same?* **no** Point to the written word *ate* or *eight* and have students demonstrate understanding by either displaying eight fingers or pantomiming eating.

Teacher Notes:

NAME _____ DATE _____

Lesson 9 Word Web
Ways to Make 8

Trace the math words. Draw a number story in each rectangle that shows the meaning of *in all*. Complete each sentence.

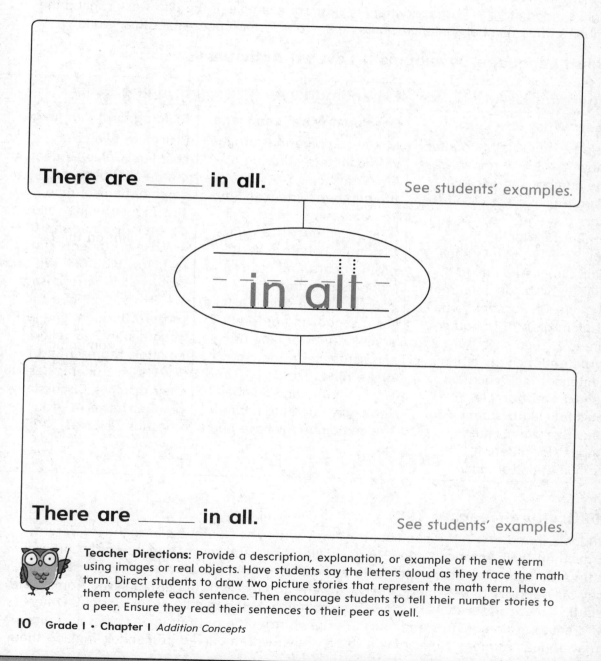

There are _____ **in all.**

See students' examples.

in all

There are _____ **in all.**

See students' examples.

Teacher Directions: Provide a description, explanation, or example of the new term using images or real objects. Have students say the letters aloud as they trace the math term. Direct students to draw two picture stories that represent the math term. Have them complete each sentence. Then encourage students to tell their number stories to a peer. Ensure they read their sentences to their peer as well.

Lesson 10 Ways to Make 9

English Learner Instructional Strategy

Sensory Support: Manipulatives

Distribute nine small sticky notes, one crayon and a write-on/wipe-off board to each pair of students. Have students draw one star on each sticky note. Assign each pair a number from zero to nine. The number the pair is assigned is the number of stars they should shade with the crayon. Once each pair has shaded their stars, have them place their sticky notes on the write-on/wipe-off board: shaded stars in one group and unshaded stars in another group. Then have them write a number sentence on their board using this sentence frame: <u>shaded stars</u> + <u>unshaded stars</u> = <u>sum</u>. For example, if the pair was assigned the number 3, their number sentence would be $3 + 6 = 9$. Have pairs share their work.

English Language Development Leveled Activities

Emerging Level	Expanding Level	Bridging Level
Show What You Know Using two colors of counters, show $4 + 5 = 9$ and write the number sentence. Say, *Four plus five equals nine. I add four and five. The sum is nine.* Distribute nine counters in a pile to each pair. Direct students to quickly grab counters from the pile. For example, one student grabs 5 the other grabs 4. Then have pairs write and recite two number sentences their counters reflect. For example, **Five plus four equals nine. Four plus five equals nine.** Repeat for different combinations of nine.	**Developing Oral Language** Model combinations of nine with counters, showing $4 + 5 = 9$ and $3 + 6 = 9$. Write the number sentences. Say, *To make nine, I add four plus five.* Emphasize *add.* Distribute counters. Write 9 on the board and have students count nine counters. Have students take turns suggesting different combinations of nine using this sentence frame, **To make nine, I add ____ and ____.** Write each number sentence. Continue until all the combinations have been given.	**Building Oral Language** Distribute two color counters. Guide students to create a display of combinations of nine, starting with nine yellow counters to represent $0 + 9 = 9$. Have students start the next row with one red and eight yellow to represent $1 + 8 = 9$. Continue the pattern through nine red counters and no yellow counters representing $9 + 0 = 9$. Have students describe their displays. Discuss any number patterns. Ask questions such as, *Which row shows $3 + 6 = 9$?*

Multicultural Teacher Tip

Students from Latin American countries may write their numbers in slightly different forms than their American peers. In particular, ones and sevens can be easily confused. Latin American ones are written with a short horizontal line at top, and at first glance, they may appear to be American-style sevens. To clearly distinguish a seven from a one, a Latin American seven will include a cross-hatch at the middle of the upright. Other differences to note: Eights and fours are drawn from the bottom up. As a result, fours at times appear as nines. Nines may also be drawn with a curved descender, making them look like lowercase "g"s.

NAME _____ DATE _____

Lesson 10 Vocabulary Definition Map
Ways to Make 9

Use the definition map to write what the math word means and tell what the word is like. Write or draw a math example. Share your examples with a classmate.

My Math Word:

add

What It Means:

To join together sets to find the total or sum.

What It Is Like:

I can add ___numbers___ like this: $3 + 6 = 9$.

When I add, I find the ___sum___.

There are many ___ways___ to make a sum of 9.

My Math Example:

See students' examples.

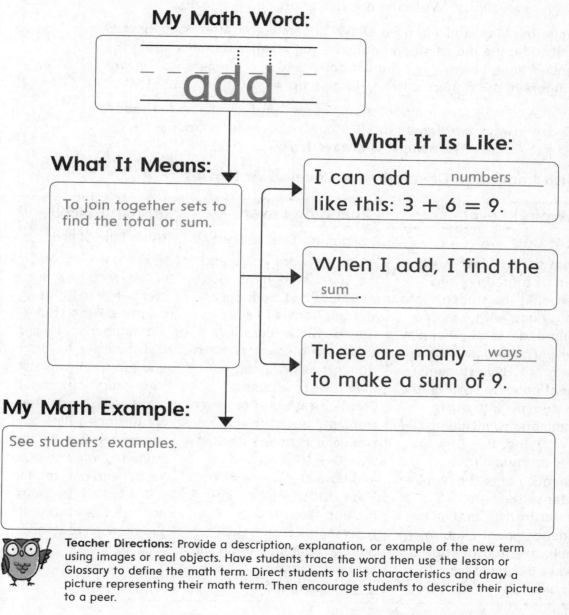

Teacher Directions: Provide a description, explanation, or example of the new term using images or real objects. Have students trace the word then use the lesson or Glossary to define the math term. Direct students to list characteristics and draw a picture representing their math term. Then encourage students to describe their picture to a peer.

Grade 1 • **Chapter 1** *Addition Concepts* **11**

Lesson 11 Ways to Make 10

English Learner Instructional Strategy

Sensory Support: Physical Activities

Before the Explore and Explain activity, hold up both hands with all ten fingers up in the air. Have students do the same. Say, *I have zero fingers down and ten fingers up. I have ten fingers in all.* Write the number sentence $0 + 10 = 10$.

Put down one finger, so that you have nine fingers up in the air. Have students do the same. Say, *I have one finger down and nine fingers up. I have ten fingers in all.* Write the number sentence $1 + 9 = 10$.

Continue this way until you have all ten fingers down. After each, have students describe the situation and write the number sentence using the sentence frame: **I have _____ fingers down and _____ fingers up. I have ten fingers in all. fingers down + fingers up = 10.**

Ask, *What part of the situation of our fingers always remains the same?* **There are always ten fingers in all.** *What part of the number sentences remains the same?* **The sum. It is always ten.**

English Language Development Leveled Activities

Emerging Level	Expanding Level	Bridging Level
Act It Out	**Developing Oral Language**	**Think-Pair-Share**
Select ten students. Position them as a group of ten. Then, split the students into two groups, with 4 in one group and 6 in another. Say, *Four plus six equals ten. The sum is ten.* Have students repeat chorally. Position the ten students into other combinations, including $0 + 10 = 10$ and $10 + 0 = 10$. After positioning the students, name the number sentence they are demonstrating, and have students repeat chorally. Emphasize that the sum is always the same for all combinations of ten.	Write the word *equal* and the Spanish cognate, *igual*. Distribute two color counters and ten-frames to each student. Model creating arrays using counters. Have students follow along on their own ten-frames. Create arrays that represent combinations of ten using two color counters starting with $0 + 10 = 10$ and ending with $10 + 0 = 10$. Have students take turns reciting the array representations of each number sentence using this sentence frame: _____ **plus** _____ ***equals* ten**	Model pairs of combinations of ten, such as $4 + 6 = 10$ and $6 + 4 = 10$, using counters. Have students say the number sentences. Distribute two color counters and ten-frames to students. Have pairs work together to make tens. If one student creates one yellow and nine red for $1 + 9 = 10$, the other student should create nine yellow and one red for $9 + 1 = 10$. Have pairs confirm the sums and write the number sentences.

Teacher Notes:

NAME _____ DATE _____

Lesson II Vocabulary Sentence Frames
Ways to Make 10

The math words in the word bank are for the sentences below. Write the words that fit in each sentence on the blank lines.

Word Bank		
equals	plus	sum

I. The answer to an addition problem is the _sum_.
    2 + 4 = 6
            ↑

2. Eight _plus_ two equals ten.

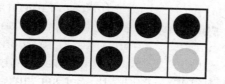

3. One plus nine _equals_ ten.

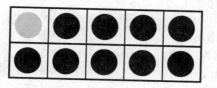

Teacher Directions: Provide a description, explanation, or example of the each term using images or real objects. Read each sentence frame and have students echo read. Direct students to write the correct term in each blank. Then encourage students to read each sentence to a peer.

Lesson 12 Finding Missing Parts of 10

English Learner Instructional Strategy

Sensory Support: Tiered Questions

Create a scene using construction paper or colored chalk/markers. Your scene should include a blue pond and green grass surrounding the pond similar to the Explore and Explain lesson. Create ten frogs, on separate sticky notes. Place 8 frogs in the water and 2 in the grass.

Gesture to the scene and say, *There are ten frogs in all. Eight frogs are in the pond. The rest are in the grass.* Write 8 + = 10.

Ask, *How many frogs are in the pond?* Have students count each frog as you point to 8 frogs in the pond. **8** Ask, *How many frogs are there in all?* Have students count on from 8 as you point to the 2 frogs in the grass. **10** Ask, *How many frogs are on the grass?* Have students count each frog that is in the grass. **2**

Gesture to the number sentence and ask, *What is the missing part?* **2**

English Language Development Leveled Activities

Emerging Level	Expanding Level	Bridging Level
Act It Out	**Show What You Know**	**Developing Oral Language**
Have students hold up ten fingers and count in unison. Say, *Ten is the whole.* Show three fingers on one hand and four on the other. Ask, *What part is missing?* Count the folded fingers and say, *Three are missing.* Continue showing other combinations. Ask, *What is the whole?* and *What part is missing?* Have students respond using this sentence frame: **The whole is ten and _____ are missing.** As students become confident, allow them to create the finger models for their classmates to solve.	Display a ten-frame filled with one color of counters. Say, *This is the whole ten.* Next, remove three counters. Ask, *What was the whole?* **ten** *What part is missing?* **three** Distribute ten-frames and counters to pairs of students. Have students in each pair alternate removing different numbers of counters from a filled ten-frame and describe it using the sentence frame: **The whole is ten. The part missing is _____.** Have pairs share their findings with the whole group.	Divide students into pairs. Have manipulatives such as counters available if needed. Distribute number cards from 0 to 10 to each student. Have one student hold up a card. Have the other student read the number, and then identify the card that creates a whole ten. Have students describe the whole and parts using the sentence frame: The whole is ten. **The parts are _____ and _____.** Continue until students have used all cards. Then have them share their experience with the whole group.

Teacher Notes:

NAME _____

DATE _____

Lesson 12 Word Identification
Find Missing Parts of 10

Match each term to a picture.

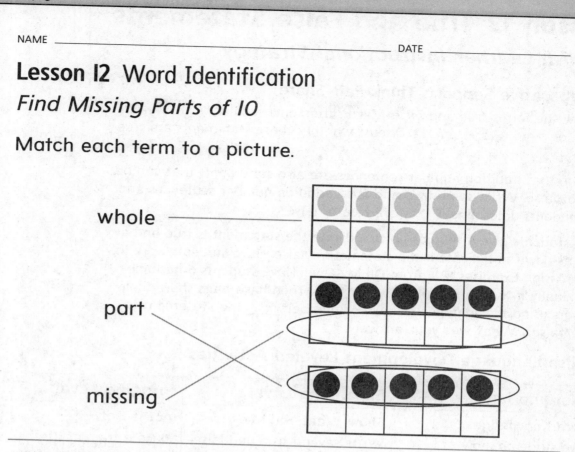

whole

part

missing

Write the correct term from above for each sentence on the blank lines.

● Part	○ Part
7	3
Whole	
10	

The ___whole___ is 10.

7 is a ___part___.

The ___missing___ part is 3.

Teacher Directions: Review the terms using manipulatives, such as counters or pennies. Have students say each word and then draw a line to match the word to its matching picture. Prompt students to describe the part-part-whole mat. Then have them write the corresponding terms in the sentences. Encourage students to read the sentences to a peer.

Grade 1 · Chapter 1 *Addition Concepts* **13**

Lesson 13 True and False Statements
English Learner Instructional Strategy

Collaborative Support: Think-Pair-Share

Review the terms *true* and *false*. Make statements such as, *I am wearing a hat.* or *I am wearing shoes.* Discuss whether these statements are true or false.

Explain that addition number sentences are also statements that can be true or false. Write some true and false addition number sentences and ask students determine if they are true or false.

Have students give a thumbs-up to indicate the statement is true and a thumbs-down to indicate if the statement is false. Have students work in pairs. Assign Exercises 5–16 from On My Own. Have students determine individually if the statement is true or false. Then have pairs share their answers using the sentence frame: **For number _____ the statement is (true/false). What was your answer?**

English Language Development Leveled Activities

Emerging Level	Expanding Level	Bridging Level
Word Knowledge	**Show What You Know**	**Act it Out**
Have students show thumbs-up for true and thumbs-down for false. Select 7 students. Place 3 in one group and 4 in another. Say, *I see three plus four.* **thumbs-up** Write and say, *true.* Say, *Three plus four equals seven. Three students plus four students equals seven students. This is true.* Emphasize true. Next, select 1 student. Place 3 in one group and 1 in another. Say, *I see three plus four.* **thumbs-down** write and say, *false.* Say, *I see three plus one. This statement is true.*	Write several true and false addition sentences on the board. Have students come to the board one at a time. Students should choose a number sentence and read it aloud. Have each student circle the statement if it is true or put an X through it if the statement is false. If the statement is false, have students explain why it is false. Discuss with students what manipulatives they could use to determine whether a statement is true or false. Have students model their thinking.	Place a train of 10 connecting cubes in one side of a bucket balance. Count a group of 3 cubes and a group of 4 cubes. Say, *Three plus four equals ten. Is this true or false?* Connect the groups and place in the other side of the balance. Say, *No, three plus four does not equal ten. The number sentence is false.* Allow students to combine connecting cubes to model other true and false math facts. If false, have students correct it to make it true.

Teacher Notes:

NAME _____ DATE _____

Lesson 13 Note Taking
True and False Statements

Read the question. Write words you need help with. Use your lesson to write your Cornell notes. Write or draw math examples to explain your thinking.

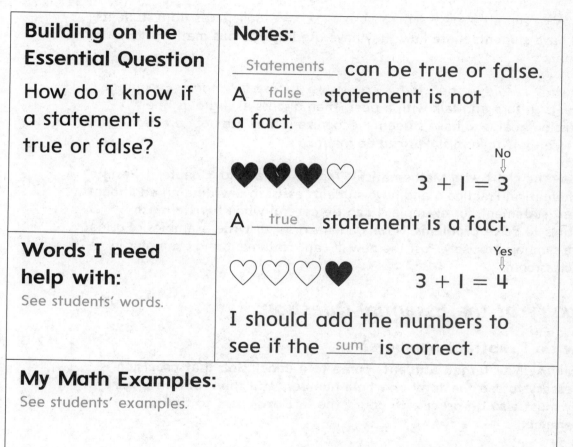

Building on the Essential Question

How do I know if a statement is true or false?

Words I need help with:

See students' words.

Notes:

Statements can be true or false.

A _false_ statement is not a fact.

♥ ♥ ♥ ♡   No
⇩
$3 + 1 = 3$

A _true_ statement is a fact.

♡ ♡ ♡ ♥   Yes
⇩
$3 + 1 = 4$

I should add the numbers to see if the _sum_ is correct.

My Math Examples:

See students' examples.

Teacher Directions: Read the Building on the Essential Question and have students list words/phrases they need assistance with. Provide descriptions, explanations, or examples of the terms using images or real objects. Read each sentence frame and have students write the appropriate terms. Have students read their notes aloud. Direct students to draw a picture representing the question. Then encourage students to describe their picture to a peer.

Chapter 2 Subtraction Concepts

What's the Math in This Chapter?

Mathematical Practice 6: Attend to precision

Write: $3 = 1 + 4$ on the board purposely placing symbols in the wrong positions. Ask students to look at the addition number sentence and raise their hand if they see anything wrong with it. Allow time for students to discuss their observations. Have a student volunteer erase the symbols and say, *I need to be **precise** or careful when I write symbols.* Then have the volunteer write the addition and equals sign correctly in the addition sentence.

Ask, Have you used math symbols like (+, −, =)? Solicit **Yes** from students, then have students share how they have used the various math symbols thus far.

Ask, *Why should we be careful when we write math symbols?* Encourage students to turn and talk with a peer. Then discuss as a group. The discussion goal is to have students recognize if they are not "precise/clear/careful" then mistakes can be made.

Review the chart with Mathematical Practice 6 made for Chapter 1. Restate Mathematical Practice 6 and have students assist in rewriting an additional "I can" statement, for example: **I can be careful when I write math symbols to solve problems.** Have students draw or write examples of using math symbols precisely. Post the new "I can" statement and examples in the classroom.

Inquiry of the Essential Question:

How do I subtract numbers?

Inquiry Activity Target: **Students come to a conclusion that precision is necessary to see the total or whole number, and then remove a part. They must also be precise to count the left over part to find the difference.**

As an introduction to the chapter, present the Essential Question to students. The inquiry graphic organizer will offer opportunities for students to observe, make inferences, and apply prior knowledge of joining parts to make a whole representing the Essential Question. As they investigate, encourage students to draw, write, and collaborate with peers to demonstrate their observations and thinking. Then have students present additional questions they may have to a peer to extend discussions. Regroup students and restate Mathematical Practice 6 and the Essential Question. Pose questions to reflect on what has been learned to guide students in making connections between the Mathematical Practice and the Essential Question.

NAME _____ DATE _____

Chapter 2 Subtraction Concepts
Inquiry of the Essential Question:

How do I subtract numbers?

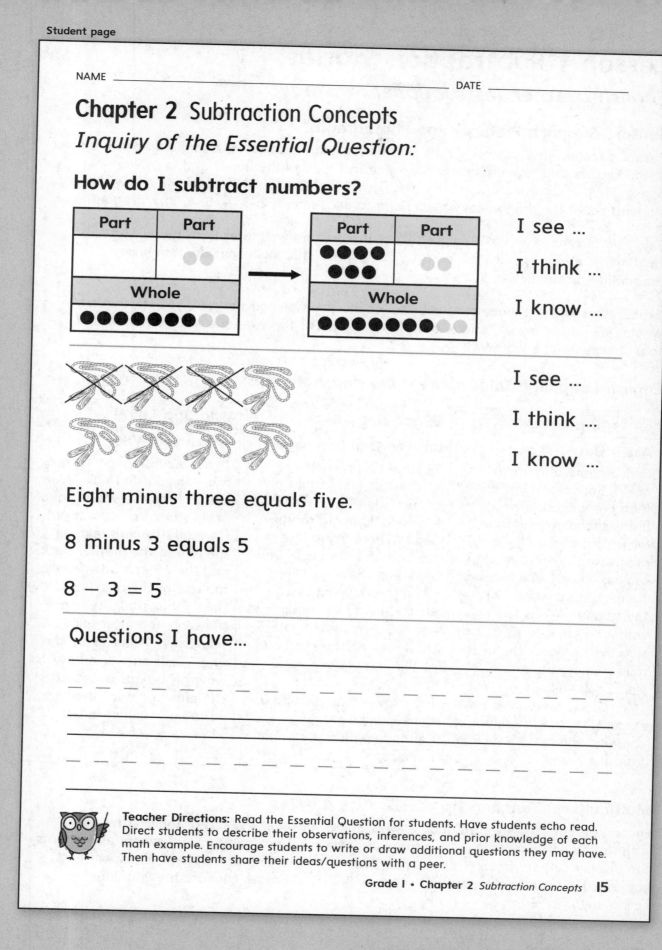

I see ...

I think ...

I know ...

I see ...

I think ...

I know ...

Eight minus three equals five.

8 minus 3 equals 5

8 − 3 = 5

Questions I have...

_____

_ _

_____

_ _

_____

Teacher Directions: Read the Essential Question for students. Have students echo read. Direct students to describe their observations, inferences, and prior knowledge of each math example. Encourage students to write or draw additional questions they may have. Then have students share their ideas/questions with a peer.

Grade 1 • **Chapter 2** *Subtraction Concepts* 15

Lesson 1 Subtraction Stories
English Learner Instructional Strategy

Sensory Support: Pictures and Photographs

Create a scene using a large picture of a flower and 7 pictures of a single dragonfly similar to the Explore and Explain lesson. Place 7 dragonflies on the flower and say, *There are seven dragonflies sitting on a flower.* Ask, *How many dragonflies are on the flower?* Have students count the dragonflies as you point to each one. **1, 2, 3, 4, 5, 6, 7; 7 dragonflies.**

Say, *Two of them fly away.* Remove two of the dragonfly pictures from the flower. Ask, *How many dragonflies are left on the flower?* Have students count the remaining dragonflies on the flower. **5**

Explain that you just modeled a subtraction story. When you subtract, you start with a whole, take away a part, and another part is left. For this example, you started with 7, took 2 away, and 5 were left.

English Language Development Leveled Activities

Emerging Level	Expanding Level	Bridging Level
Act It Out Play a game of musical chairs. Set up one less chair than the number of students in the group. Arrange the chairs in a circle. Play a progressive take-away song, such as, "Five Little Monkeys Jumping on the Bed." Allow the student who is left without a chair to start and stop the music for the next round. Before the music starts, say, *Take away one chair.* Stress *take away.* Have students repeat the sentence chorally, and remove a chair from the game.	**Building Oral Language** Teach a progressive subtraction song using one of the following: "Five Little Monkeys Jumping on the Bed," "There Were Five in the Bed and the Little One Said, Roll Over," or "Five Little Ducks Went Out to Play." Select five students to act out the song. Have students sing the song and at the end of each verse, stop and say, *Take away one.* Have students repeat the sentence chorally. Continue until one student remains.	**Watch and Write** Teach a subtraction song like, "Five Little Ducks Went Out to Play." Select a group of six students to act it out. Have one student take on the role of the mother duck and the other students take the role of the ducklings. Have other students sing the song and after each verse, say, **"Take away the one."** Distribute write-on/wipe-off boards to seated students and have them write how many are left after each verse.

Multicultural Teacher Tip

Encourage ELs to share traditions, stories, songs, or other aspects of their native culture with the other students in class. You might even create a "culture wall" where all students can display cultural items. This will help create a classroom atmosphere of respect and appreciation for all cultures, and in turn, will create a more comfortable learning environment for ELs.

NAME _____ DATE _____

Lesson I Word Web
Subtraction Stories

Trace the math words. Draw a number story in each rectangle that shows the meaning of *are left*. Complete the sentences about each story.

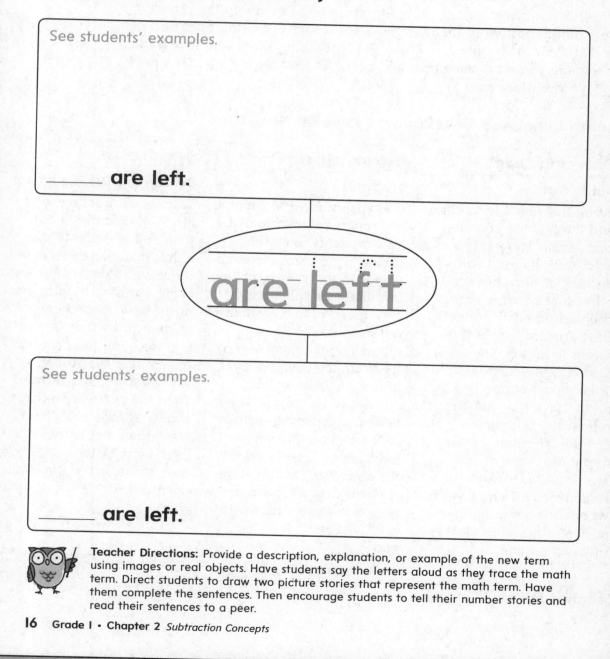

See students' examples.

_____ **are left.**

are left

See students' examples.

_____ **are left.**

Teacher Directions: Provide a description, explanation, or example of the new term using images or real objects. Have students say the letters aloud as they trace the math term. Direct students to draw two picture stories that represent the math term. Have them complete the sentences. Then encourage students to tell their number stories and read their sentences to a peer.

Lesson 2 Model Subtraction

English Learner Instructional Strategy

Graphic Support: Graphic Organizers

Use a large presentation part-part-whole mat (or draw one) and pictures to demonstrate the lesson. Have pictures of 10 toys ready for the Model Subtraction activity. Say, *There are ten toys.* Place the pictures of the toys on the bottom "whole" part of the mat and write the number 10. Say, *A boy takes eight toys out of the box.* Move 8 of the toys to the top left part of the mat. Have students count as you move each one. Write, 8. Say, *I want to know how many toys are left in the box.* Gesture to the remaining 2 toys on the bottom of the mat. Move the 2 toys to the top right part of the mat. Count them off: *1, 2* Write, 2. Model using these sentence frames for students to report the whole and parts: **The whole is _____. One part is _____. The other part is _____.**

English Language Development Leveled Activities

Emerging Level	Expanding Level	Bridging Level
Act It Out	**Synthesis**	**Developing Oral Language**
Write the word *subtraction* and the Spanish cognate, *sustracción.* Have students wash their hands. Distribute a snack of 10 raisins (or other snack that is easily counted) to each student. Have students count the raisins in unison. Say *I will subtract one raisin. Yum!* Then model eating one raisin. Have the students repeat the sentence chorally and then eat one raisin. Ask, *How many raisins are left?* Have students count and answer. 9 Continue subtracting one or two raisins until all the snacks are gone.	Give each student 10 counters and a part-part-whole mat. Have students count the counters and place them red-side-up in the *Whole* section. Distribute one spinner to each group of four. Students will spin and model subtracting the number they spin from the counters, by turning the appropriate number of counters yellow-side-up and placing them in a *Part* section. The leftover red counters are placed in the other *Part* section. Have students describe using this sentence frame: **The whole is 10. The parts are _____ and _____.**	Teach a subtraction song like, "The Farmer in the Dell." Have students join hands in a circle and count off. Have students walk in a circle and sing the first part of the song. Select one student to be the "farmer" and step into the center of the circle. Have students say, **Subtract the farmer.** Then the circle of students counts off again, without the farmer. Repeat with each verse. After each verse, have those left in the circle count off.

Teacher Notes:

NAME _____ DATE _____

Lesson 2 Four-Square Vocabulary
Model Subtraction

Trace the word. Write the definition for *subtract*.
Write what the word means, draw a picture, and
write your own sentence using the word.

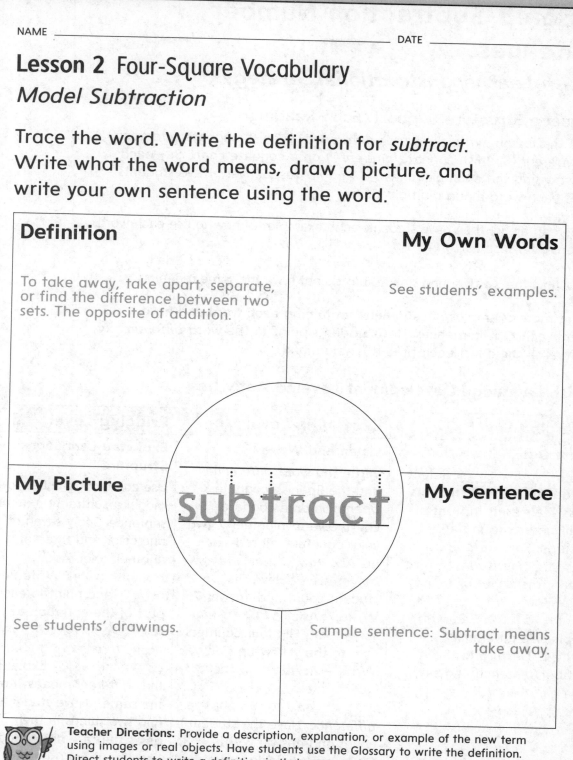

Definition

To take away, take apart, separate, or find the difference between two sets. The opposite of addition.

My Own Words

See students' examples.

My Picture

See students' drawings.

subtract

My Sentence

Sample sentence: Subtract means take away.

Teacher Directions: Provide a description, explanation, or example of the new term using images or real objects. Have students use the Glossary to write the definition. Direct students to write a definition in their own words and draw a picture representing their math term. Have students write a sentence using the term. Then encourage students to read their sentence to a peer.

Grade 1 • **Chapter 2** *Subtraction Concepts* **17**

Lesson 3 Subtraction Number Sentences

English Learner Instructional Strategy

Language Structure Support: Echo Reading

Before the lesson, write the words *difference* (Spanish cognate *diferencia*) and *different* (Spanish cognate *diferente*.) on a cognate chart. Say each word stressing the ending sounds /s/ and /t/ respectively. Have students repeat each word aloud multiple times.

Write, read aloud, then have students echo examples of how to use each word in a sentence; such as *Three minus two equals one. The difference is one.* and *The numbers three and four are not the same, they are different.* Briefly discuss how the words sound similar, but do not have the same meaning.

Present more examples of sentences containing each word and have students echo the sentences. Have students point to the word *difference* or *different* on the chart each time it is said aloud.

English Language Development Leveled Activities

Emerging Level	Expanding Level	Bridging Level
Act It Out	**Listen and Write**	**Exploring Language Structure**
Write $7 - 5 =$ ____. Select 7 students. Have 5 students cross their arms in front of their faces creating an X-shape. Say, *Here are seven students; five students have been crossed out or taken away. The number that remains is the difference.* Emphasize difference. Write 2. Gesture to the number sentence and say, *Five minus, or take away, three equals two. The difference is two. This is a subtraction number sentence.* Repeat with other numbers. Have students write the difference on write-on/wipe-off boards.	Write the word *difference* and the Spanish cognate, *diferencia* on a cognate chart. Give a volunteer 7 two color counters, all red-side up. Ask, *How many counters do you have?* Have the student count and answer 7. Write 7. Ask, *What if I take away four?* Flip four counters over to the yellow-up side. Write − 4. Ask, *How many red counters remain? What is the difference?* Emphasize *difference*. Have the student count and answer. Write = 3. Give pairs two color counters and have them repeat the activity using different numbers.	Use connecting cubes to model a subtraction number sentence. Show seven of one color and three of another color. Ask, *What is the difference?* Write $7 - 3 =$ ____ and point to each part of the equation as you say, *Seven minus, or take away, three equals four. The difference is four.* Explain that *different* means "not the same." Have students find two things in their desks that are different and describe them using this sentence frame: **These are different because ____. They differ.**

Teacher Notes:

NAME _____ DATE _____

Lesson 3 Word Identification
Subtraction Number Sentences

Match each term to a symbol or example.

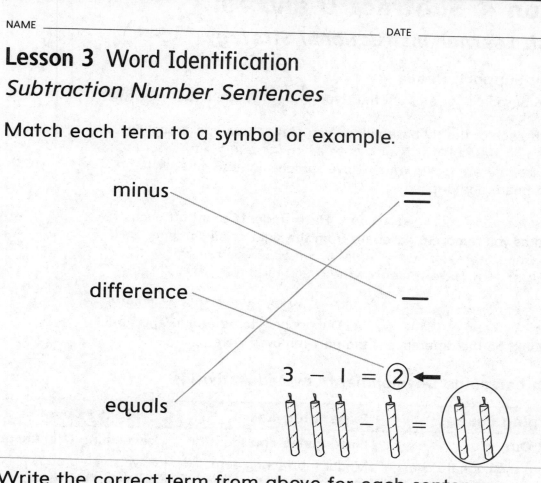

Write the correct term from above for each sentence on the blank lines.

subtraction number sentence
$$3 - 1 = 2$$

Three __minus__ one __equals__ two.

The __difference__ is two.

Teacher Directions: Review the terms using manipulatives, such as counters or pennies. Have students say each word and then draw a line to match the word to its symbol or example. Direct students say the subtraction number sentence and then write the corresponding terms in the sentences. Encourage students to read the sentences to a peer.

Lesson 4 Subtract 0 and All
English Learner Instructional Strategy

Sensory Support: Realia

Collect 6 baseballs (or 6 of another type of ball) to use in this activity.

Count off each of the six baseballs as you place them in a group. Say, *A team had six baseballs at their game. Write the number to describe how many baseballs are in the whole.* Have students write 6 on a write-on/wipe-off board and display.

Say, *The team lost 6 of the baseballs.* Have students count off each baseball as you remove 6 baseballs from the sight of all students. *How many baseballs are left?* **none** *Write a number to describe how many baseballs are left.* Have students write 0 and display.

Ask, *What number describes the part removed?* **6** *Was the part removed all, some, or none of the whole?* **all** Discuss how to determine this and what would be the difference if the part removed were 0.

English Language Development Leveled Activities

Emerging Level	Expanding Level	Bridging Level
Act It Out Refer to the cognate chart and review *zero (cero)*. Teach a progressive subtraction song, such as, "Five Little Ducks Went Out to Play." Select 5 students to act out the song as you sing it. At the song's conclusion, ask, *How many ducks are left?* **none** Gesture to *zero* and say, *zero.* Have students repeat. Write the number sentence $5 - 5 = 0$. Have students read the sentence with you as you point to each part. Repeat with other students acting out the song.	**Look, Listen and Identify** Model subtracting zero using counters. Start by showing five counters. Have students chorally count with you to find the total number of counters: **1, 2, 3, 4, 5.** Say, *I'm going to take away zero counters from the five counters.* Gesture with your hand as if you were taking away some counters from the group. Ask, *Now how many counters?* **5** Have students chorally count with you to find the total number of counters. Write the subtraction number sentence $5 - 0 = 5$.	**Developing Oral Language** Have small groups play a game with cards numbered from 1–10 (four cards with each number). Each player draws five cards. The rest lay in a draw pile. Each player sets any matches (two cards with the same number) face-up. Players must describe the match using the following sentence frame: ____ **minus** ____ **equals zero.** Players take turns drawing a card and draw again if a match is made. The object is to be the first to match all of his or her cards and have zero cards left.

Teacher Notes:

NAME _____ DATE _____

Lesson 4 Note Taking
Subtract 0 and All

Read the question. Write words you need help with. Use your lesson to write your Cornell notes. Write or draw math examples to explain your thinking.

Building on the Essential Question	**Notes:**
How can I subtract 0? How can I subtract all?	I know that *subtract* means "_take_ _away_." If I subtract all, I will have _0_ left. ◯◯◯ – ◯◯◯ = 0
Words I need help with: See students' words.	If I subtract 0 from a number, I will have the _same_ _number_ left. ◯◯◯◯ – 0 = ◯◯◯◯
My Math Examples: See students' examples.	

Teacher Directions: Read the Building on the Essential Question and have students list words/phrases they need assistance with. Provide descriptions, explanations, or examples of the terms using images or real objects. Read each sentence frame and have students write the appropriate terms. Have students read their notes aloud. Direct students to draw a picture representing the question. Then encourage students to describe their picture to a peer.

Grade 1 · **Chapter 2** *Subtraction Concepts* **19**

Lesson 5 Vertical Subtraction

English Learner Instructional Strategy

Vocabulary Support: Draw Visual Examples

Write the following sentence horizontally and vertically (write one word on each line in a vertical list), *My favorite color is red.* Ask a volunteer to read the horizontal sentence and another volunteer to read the vertical sentence while the other students have their eyes closed. Ask, *Were the two sentences you heard the same?* **yes**

Now write the following subtraction number sentence horizontally and vertically, $3 - 1 = 2$. Ask a volunteer to read the horizontal sentence and another volunteer to read the vertical sentence while the other students keep their eyes closed. Ask, *Were the two number sentences you heard the same?* **yes**

Have all students open their eyes and look at what is written. Discuss how the sentences are the same whether written horizontal or vertical.

English Language Development Leveled Activities

Emerging Level	Expanding Level	Bridging Level
Word Recognition	**Think-Pair-Share**	**Developing Oral Language**
Write $3 - 2 = 1$ on the board horizontally and say, *Three minus two equals one. The difference is one.* Model the equation with two colors of connecting cubes. Write $3 - 2 = 1$ in vertical form. Read and model it again. Say, *Each subtraction number sentence is the same.* Write different subtraction number sentences on the board, both vertically and horizontally. Have students read them aloud in unison. Ask, *What is the difference?* Have them answer with a vertical or horizontal gesture, or use this sentence frame, **The difference is _____.**	Write a subtraction number sentence, such as $4 - 2 = 2$, horizontally and vertically. Read and model each using two color counters. Explain that it is the same written either way. Distribute two color counters to pairs of students, with each partner taking away (flipping over) counters. Have pairs model a subtraction number sentence while saying the number sentence. Have pairs write the number sentence in horizontal and vertical form in their math journals and then share with another pair/small group of students.	Write a subtraction number sentence, such as $4 - 2 = 2$, in horizontal and vertical forms. Have students read each number sentence aloud. Explain that it is the same written either way. Provide pairs of students a subtraction number sentence, such as: $5 - 2 = 3$. On write-on/wipe-off boards, have one partner write the sentence horizontally and the other partner vertically. Have pairs share their subtraction sentences. Ask, *What is the difference?* Have students answer using this sentence frame, **The difference is _____.**

Teacher Notes:

NAME _____ DATE _____

Lesson 5 Concept Web
Vertical Subtraction

Write *vertical* in the center oval. Draw lines to match the vertical items to the word *vertical*.

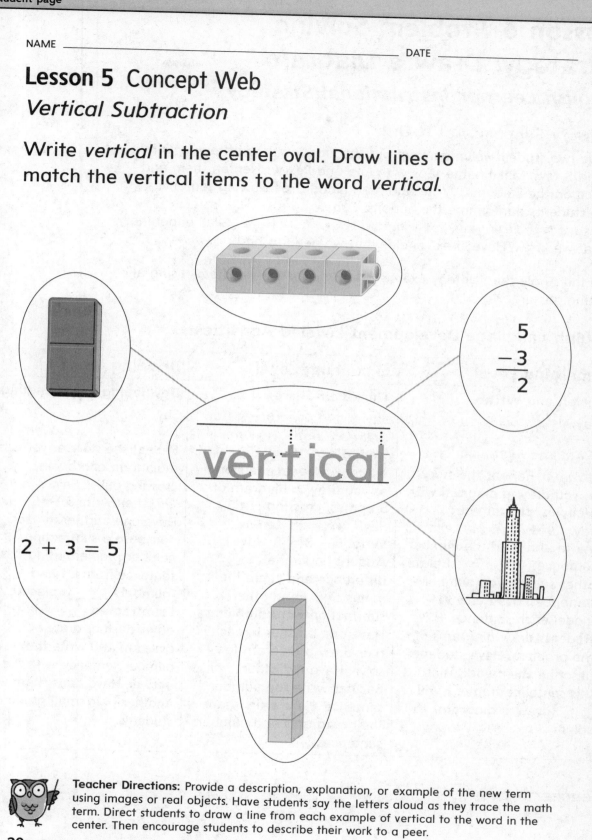

Teacher Directions: Provide a description, explanation, or example of the new term using images or real objects. Have students say the letters aloud as they trace the math term. Direct students to draw a line from each example of vertical to the word in the center. Then encourage students to describe their work to a peer.

20   Grade 1 · Chapter 2 *Subtraction Concepts*

Lesson 6 Problem Solving Strategy: Draw a Diagram

English Learner Instructional Strategy

Sensory Support: Act It Out

Have two student volunteers act out the Practice the Strategy activity using 8 toys. Make name tags with Lila and Rex written on them and place on the students. Write and read aloud the following sentences as the students pantomime the actions: *Lila has 8 toys. She lets Rex play with 3 of the toys. How many toys does Lila have left?* Have "Lila" underline what we know. Have "Rex" circle what we need to find.

For the Apply the Strategy Exercises 1–3, review the word *eats* using an eating gesture.

English Language Development Leveled Activities

Emerging Level	Expanding Level	Bridging Level
Listen and Write Say, *I had four buttons. I lost two. How many do I have left? I will draw a diagram.* Repeat the story as you draw a diagram with 4 circles, crossing out 2. Say, *I have two buttons left.* Write and read the number sentence: *4 − 2 = 2.* Relate other subtraction number sentences. Use items to model each sentence. Have students draw diagrams of the problem. Have students describe their work using this sentence frame: **I will ____ (draw a diagram) to solve.**	**Think-Pair-Share** Say, *I had five apples. Now there are three. How many were eaten? I will draw a diagram.* Repeat the story as you draw a diagram of 5 apples, crossing out 3. Say, *Two were eaten.* Write 5 − 3 = 2. Have a student underline the difference and circle the minus sign. Read other subtraction word problems. Have one partner in each pair draw a diagram to solve. Have the other partner write the number sentence. Have pairs share their diagrams and number sentences.	**Developing Oral Language** Say, *I had five books. I read two. How many are left?* Repeat the story as you draw a diagram of 5 books, crossing out 2. Say, *Three are left.* Then write 5 − 2 = 3. Have one partner in each pair pose a subtraction number problem using this sentence frame: **I had ____ (number) ____ (objects). I subtracted ____.** Have the other partner draw a diagram and write the number sentence with the answer. Have pairs share with another pair/small group of students.

Teacher Notes:

NAME _____ DATE _____

Lesson 6 Problem Solving
STRATEGY: *Draw a Diagram*

<u>Underline</u> what you know. (Circle) what you need to find. Draw a diagram to solve.

I. <u>There are **9** frogs on a tree.</u>

<u>**4** of the frogs **hop away**.</u>

How many frogs **are left** on the tree?

frog

tree

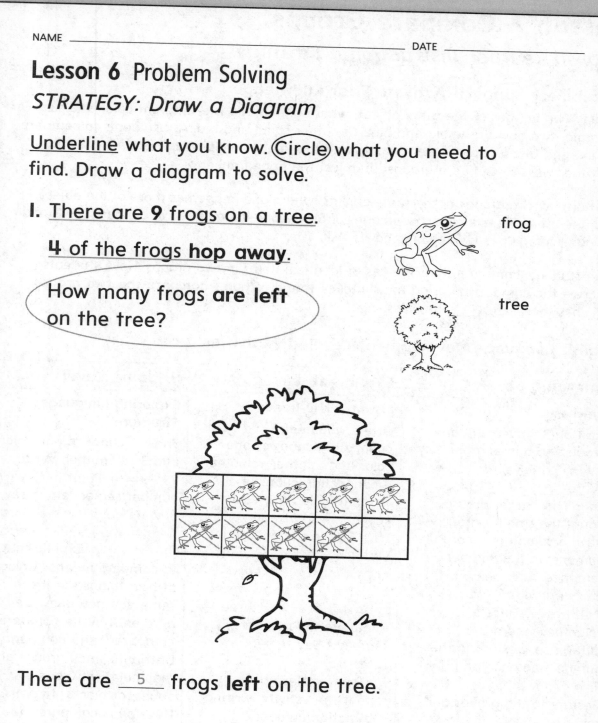

There are __5__ frogs **left** on the tree.

Teacher Directions: Provide a description, explanation, or example of the boldface terms and nouns using images or real objects. Read each sentence and have students echo read. Encourage students to use the ten-frame to diagram the problem, and then write their answer in the restated question. Have students read the answer sentence aloud.

Grade I • **Chapter 2** *Subtraction Concepts* **21**

Lesson 7 Compare Groups

English Learner Instructional Strategy

Vocabulary Support: Activate Prior Knowledge

Display two images of the same object, where one image is a different color and size. For example, two bowls: one big and blue the other small and red. Ask students to compare the images. Discuss how students compared the images (color, size). Explain that there are other ways to compare groups, than just the appearance.

Next display two groups of the same objects, where one image has more of the object than the other. For example, two images of bowls: one image has five big blue bowls and the other image has two big blue bowls. Ask, *Compare the groups; which has more bowls?*

Have students compare other images of identical items in two groups. Have students describe the comparison using the sentence frame: **When I compare the two groups, ____ has more and ____ has less.**

English Language Development Leveled Activities

Emerging Level	Expanding Level	Bridging Level
Synthesis	**Listen-Write-Read**	**Exploring Language Structure**
Position 3 students in one group and 5 in another. Say, *I will compare the groups.* Write ____ − ____ = ____. Count the larger group. Count the smaller group. Have a volunteer record the numbers in the number sentence. Ask, *How many more are in this group than that group?* Have a volunteer record the difference. Gesture to the number sentence and say, *Five minus three equals two.* Gesture to the groups and say, *There are two more in this group than that group.*	Give a volunteer 7 connecting cubes of one color and another volunteer 4 connecting cubes of a different color. Ask each, *How many cubes do you have?* Have students count. Write 7 − 4 = ____. Say, *Let's compare the number of cubes. How many more does she/he have?* **three** Complete the number sentence. Say, *I subtracted three from seven to compare the groups.* Have pairs repeat the activity, writing subtraction number sentences and using your model as a sentence frame.	Show 7 cubes of one color and 3 of another. Write 7 − 3 = 4. Point to the number sentence and say, *Seven minus three equals four. There are four more in this group.* Explain that to compare means "to look at how things are the same and how they are different." Write *compare, compared,* and *comparing.* Distribute small groups of manipulatives (counters, cubes, or blocks) to pairs. Have pairs compare the groups. Allow pairs to describe their findings to other pairs/small group of students using the various verb forms of *compare.*

Teacher Notes:

NAME _____ DATE _____

Lesson 7 Vocabulary Definition Map
Compare Groups

Use the definition map to write what the math word means and tell what the word is like. Write or draw a math example. Share your examples with a classmate.

My Math Word:

compare

What It Means:

Look at objects, shapes, or numbers and see how they are alike or different.

What It Is Like:

I can subtract to compare ___groups___.

I can tell which group has ___more/fewer___.

Equal groups have the ___same___ ___number___ of objects.

My Math Example:

See students' examples.

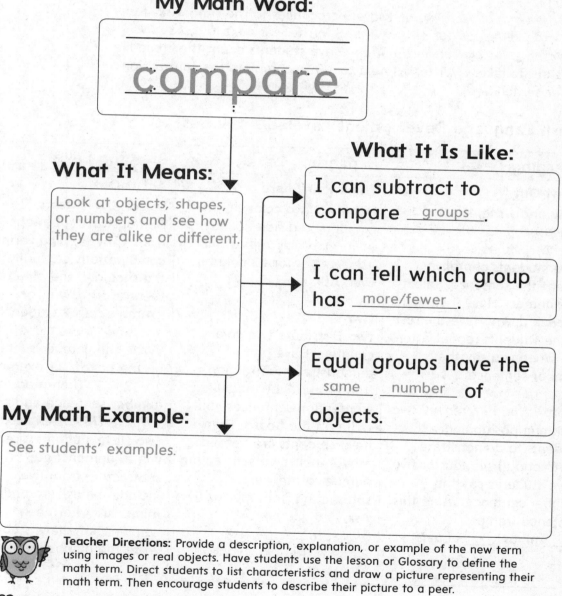

Teacher Directions: Provide a description, explanation, or example of the new term using images or real objects. Have students use the lesson or Glossary to define the math term. Direct students to list characteristics and draw a picture representing their math term. Then encourage students to describe their picture to a peer.

Lesson 8 Subtract from 4 and 5
English Learner Instructional Strategy

Sensory Support: Illustrations/Pictures

Create a scene using construction paper or colored chalk/markers. Your scene should include a blue lake and brown sand surrounding the lake similar to the Explore and Explain lesson. Create four crocodiles, on separate sticky notes. Place 4 crocodiles in the water. Gesture to the scene and say, *There are four crocodiles in a lake.* Ask, *How many crocodiles are in the lake?* **4** Have students count each crocodile as you point to them. Write 4 on the board.

Say, *Three crocodiles get out.* Move 3 crocodiles to the sand. Ask, *How many crocodiles got out of the lake?* **3** Write − 3 next to 4. Ask, *How many crocodiles are left in the lake?* Have students count the crocodile that is in the lake. **1** Write = 1 next to −3 to complete the subtraction number sentence.

English Language Development Leveled Activities

Emerging Level	Expanding Level	Bridging Level
Act It Out	**Think-Pair-Share**	**Act It Out**
Show and write 4 − 3 = 1. Say, *Four minus three equals one. The difference is one.* Have students chorally repeat the sentence. Select 5 volunteers. Have 3 students sit back down. Have the rest of the students repeat after you when you say the number sentence: *Five minus three equals two.* Ask, *What is the difference?* **two** Continue positioning students to demonstrate subtracting from four or five. Have students say the number sentence, using this sentence frame: ____ **minus** ____ **equals** ____.	Display 5 two color counters, red side up. Say, *I will model five minus three equals two.* Turn 3 counters over. Ask, *How many counters are red side up?* **two** *What is the difference?* **two** Distribute two color counters. Have pairs demonstrate subtracting from 4 and 5 by modeling subtraction sentences you write on the board. Then have students create and model their own subtraction number sentences, subtracting from four or five. Ask, *Did you prefer to model the sentences I wrote or the ones you created?*	Select 4 volunteers. Demonstrate subtracting from 4, by tapping various combinations of students on the shoulder and directing them to sit down. For example, tap 2 students and say, *Please, sit down.* Write the number sentence on the board, for example: 4 − 2 = 2. Then read the number sentence and ask, *What is the difference?* **two** Have students direct you to demonstrate subtracting from five. Include minus zero and minus all. After each combination, have students write the number sentence.

Teacher Notes:

NAME _____ DATE _____

Lesson 8 Difference Identification
Subtract from 4 and 5

Draw a line to match to the difference.

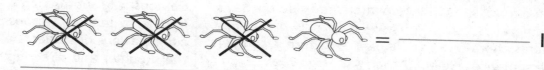

3

2

0

1

Draw a picture to show one way to subtract from 5.

See students' examples.

 Teacher Directions: Use manipulatives to show various ways to subtract from 4 and 5. Model a subtraction sentence for each and have students repeat it. Have students draw a line to match each subtraction sentence to the difference. Direct students to draw a picture representing a way to subtract from 5. Then encourage students to describe their picture to a peer.

Grade I · Chapter 2 *Subtraction Concepts* **23**

Lesson 9 Subtract from 6 and 7

English Learner Instructional Strategy

Collaborative Support: Act It Out

Select 7 students to act out the Explore and Explain activity. Say, *There are seven apes eating bananas.* Have the seven students pretend to be apes eating bananas. Say, *Three apes stopped eating bananas.* Have three of the students form their own group and stop pretending to be apes eating bananas. Ask, *How can we determine how many apes are still eating bananas?*

Gesture to all the students and ask, *How many apes were eating bananas at first?* **7** Gesture to the group of three and ask, *How many apes stopped eating bananas?* **3** Gesture to the group of four and ask, *How many apes are still eating bananas?* **4** Write 7 − 3 = 4. Rearrange the 7 students and change the story to model other ways to subtract from 7. For example, none of the apes stopped eating bananas will represent 7 − 0 = 7.

English Language Development Leveled Activities

Emerging Level	Expanding Level	Bridging Level
Act It Out	**Think-Pair-Share**	**Act It Out**
Show and write 7 − 6 = 1. Say, *Seven minus six equals one. The difference is one.* Have students chorally repeat the sentences. Select 6 volunteers. Have 1 student sit back down. Have the rest of the students repeat after you when you say the number sentence: *Six minus one equals five.* Ask, *What is the difference?* **five** Continue positioning students to demonstrate subtracting from six or seven. Have students say the number sentence, using this sentence frame: _____ **minus** _____ **equals** _____.	Display 6 two color counters, red side up. Say, *I will model six minus four equals two.* Turn 4 counters over. Ask, *How many counters are red side up?* **two** *What is the difference?* **two** Distribute two color counters. Have pairs demonstrate subtracting from 6 and 7 by modeling subtraction sentences you write on the board. Then have students create and model their own subtraction number sentences, subtracting from six or seven. Ask, *Did you prefer to model the sentences I wrote or the ones you created?*	Assemble 6 volunteers. Demonstrate subtracting from 6, by tapping various combinations of students on the shoulder and directing them to sit down. For example, tap 2 students and say, *Please, sit down.* Write the number sentence on the board, for example: 6 − 2 = 4. Then read the number sentence and ask, *What is the difference?* **four** Have students direct you to demonstrate subtracting from seven. Include minus zero and minus all. After each combination, have students write the number sentence.

Teacher Notes:

NAME _____ DATE _____

Lesson 9 Difference Identification
Subtract from 6 and 7

Match a subtraction sentence to a way to
subtract from 7.

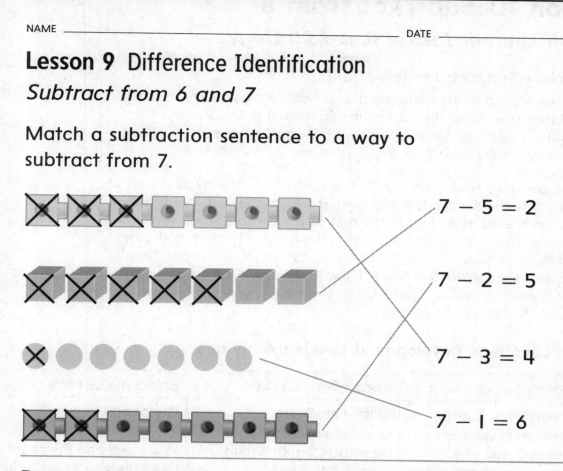

7 – 5 = 2

7 – 2 = 5

7 – 3 = 4

7 – 1 = 6

Draw a picture to show one way to subtract from 6.

See students' examples.

Teacher Directions: Use manipulatives to show various ways to subtract from 6 and 7.
Model a subtraction sentence for each and have students repeat it. Have students draw
a line to match each image to the correct subtraction sentence. Then direct students to
draw a picture representing a way to subtract from 6. Finally encourage students to
describe their picture to a peer.

Lesson 10 Subtract from 8

English Learner Instructional Strategy

Vocabulary Support: Modeled Talk

Model answering the Talk Math question. Once complete, have students remodel to a peer. Show the subtraction problem $8 - 5 = 3$ using manipulatives. Say, *Eight minus five equals three. The whole is eight. The part I know is five. The other part, which is the difference, is three.*

Display a part-part-whole mat. Write the number 8 in the "whole" part of the mat. Write the number 5 in the top left part of the mat. Write the number 3 in the top right part of the mat. Say, *I can check my answer. I know that one part is five and the other part is three.* Gesture to the two parts on the mat. Say, *I can add the two parts to find the whole.* Five plus three equals eight. Write $5 + 3 = 8$. Point to the whole on the mat and say, *The whole is equal to eight. I have checked my work and I know eight minus five equals three.*

English Language Development Leveled Activities

Emerging Level	Expanding Level	Bridging Level
Show What You Know Model different ways to subtract from eight by showing combinations of two colors of connecting cubes. With each combination, write the number sentence. For example, show and write $8 - 5 = 3$. Say, *Eight minus five equals three. The difference is three.* Write and show $8 - 4 = 4$. Ask, *What is the difference?* **four** Distribute two colors of connecting cubes to students. Have students create models as you say subtraction number sentences. After modeling, have students read the subtraction number sentence in unison.	**Think-Pair-Share** Use connecting cubes to demonstrate subtracting from eight: $8 - 7 = 1$, $8 - 0 = 8$, and so on. After each model, ask, *What is the difference?* Distribute connecting cubes and have pairs show and write the number sentences for as many ways to subtract from eight as they can. Have pairs share their sentences with another pair or the whole group. Discuss afterward any ways to subtract from eight that the groups might have missed.	**Exploring Language Structure** Write *eight* on the board. Have students count to eight and display eight using their fingers. Write *ate* on the board. Say, *I ate my breakfast.* Demonstrate ate by pantomiming eating food. Have students say, eight and ate, as you point to each word. Ask, *Do these words sound the same?* **yes** *Are these words spelled the same?* **no** Point to the written word *ate* or *eight* and have students demonstrate understanding by either displaying eight fingers or pantomiming eating.

Teacher Notes:

NAME _____ DATE _____

Lesson 10 Vocabulary Sentence Frames
Subtract from 8

The math words in the word bank are for the sentences below. Write the words that fit in each sentence on the blank lines.

Word Bank		
minus	difference	equals

1. The answer to a subtraction problem is the _____difference_____.

   8 − 5 = 3
         ↑

2. Eight minus four ___equals___ four.

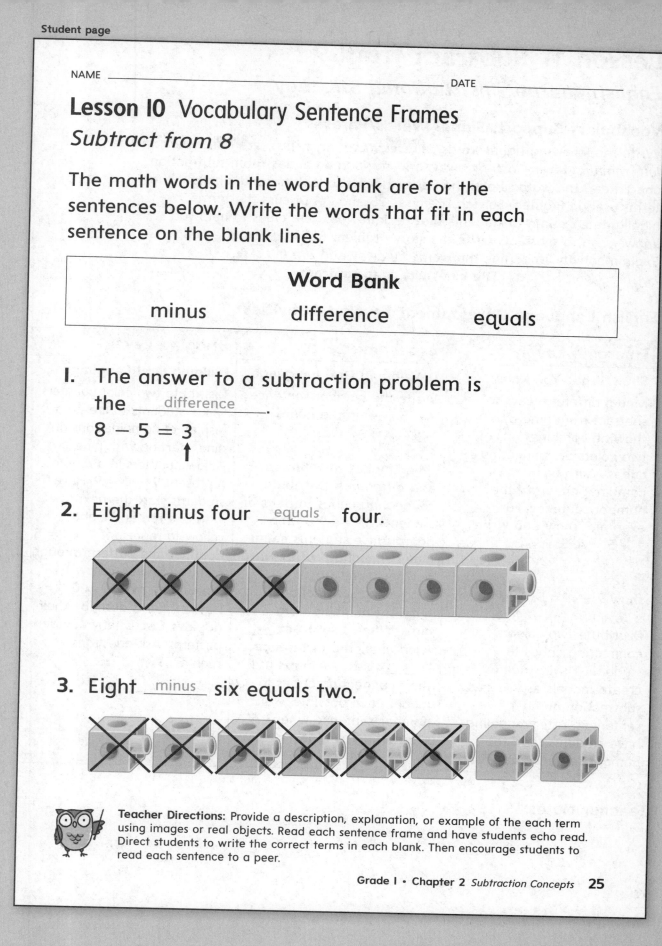

3. Eight ___minus___ six equals two.

Teacher Directions: Provide a description, explanation, or example of the each term using images or real objects. Read each sentence frame and have students echo read. Direct students to write the correct terms in each blank. Then encourage students to read each sentence to a peer.

Grade 1 • **Chapter 2** *Subtraction Concepts* **25**

Lesson 11 Subtract from 9

English Learner Instructional Strategy

Vocabulary Support: Signal Words/Phrases

Write the following signal words/phrases: *take away, less, are left, have left, remains,* fewer, *lost, fly away,* and *stopped* on a classroom subtraction chart. Read the words aloud then ask students to look for and highlight the terms used in Problem Solving Exercises 25 and 26. Discuss the terms highlighted. Explain to students that certain words are clues to help you know when to subtract. Write the story problem from the Explore and Explain activity. Underline the words *fly away* and *are left.* Ask, *What clue do these words tell us?* **The clue tells us to subtract.**

English Language Development Leveled Activities

Emerging Level	Expanding Level	Bridging Level
Show What You Know Model different ways to subtract from nine by showing combinations of two colors of connecting cubes. With each combination, write the number sentence. For example, show and write 9 − 5 = 4. Say, *Nine minus five equals four. The difference is four.* Write and show 9 − 4 = 5. Ask, *What is the difference?* **five** Distribute two colors of connecting cubes to students. Have students create models as you say subtraction number sentences. After modeling, have students read the number sentence in unison.	**Developing Oral Language** Write the number sentence 9 − 5 = 4 on the board. Say, *When I subtract five from nine, the difference is four.* Emphasize *subtract* and *difference.* Distribute connecting cubes to pairs of students. Write 9 on the board. Have students count nine connecting cubes. Have partners take turns suggesting subtracting different numbers from nine: **When I subtract _____ from nine, the difference is _____.** Have students use their connecting cubes to model each possible number sentence when subtracting by nine.	**Building Oral Language** Distribute two color counters. Guide students to create a display of combinations of nine, starting with nine yellow counters in a row to represent 9 − 0 = 9. Have students start the next row with one red and eight yellow to represent 9 − 1 = 8. Continue the pattern through nine red and no yellow representing 9 − 9 = 0. Have students describe their displays. Discuss any number patterns. Ask questions such as, *Which row shows 9 − 6 = 3?*

Teacher Notes:

NAME _____ DATE _____

Lesson II Word Web
Subtract from 9

Read the words in the word bank. Write the subtraction terms in the ovals.

Word Bank

take away   plus   difference

are left   minus   sum   in all

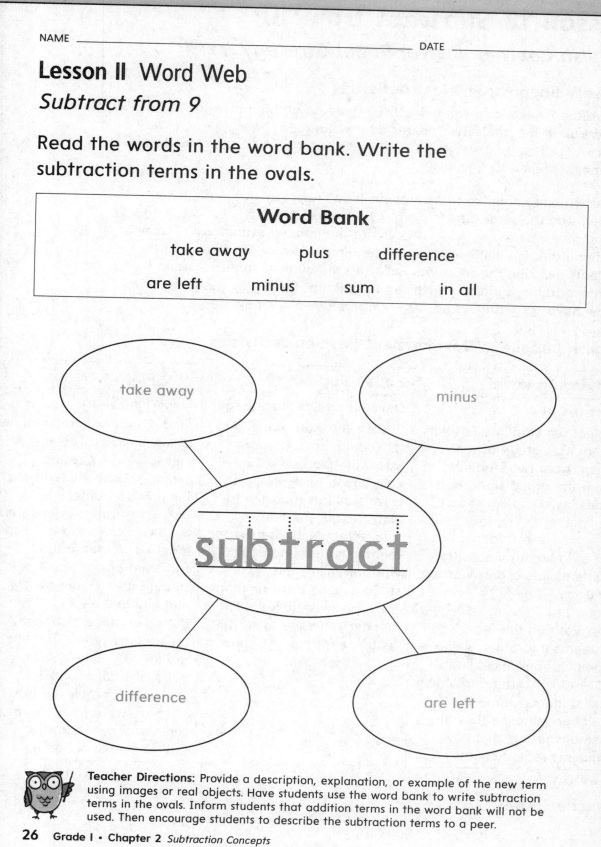

Teacher Directions: Provide a description, explanation, or example of the new term using images or real objects. Have students use the word bank to write subtraction terms in the ovals. Inform students that addition terms in the word bank will not be used. Then encourage students to describe the subtraction terms to a peer.

Lesson 12 Subtract from 10

English Learner Instructional Strategy

Sensory Support: Physical Activities

Before the Explore and Explain activity, hold up both hands with all ten fingers up in the air. Have students do the same. Say, *I show ten fingers up. I will put zero fingers down. Now I still have ten fingers up.* Write the number sentence $10 - 0 = 10$.

Put down one finger, so that you have nine fingers up in the air. Have students do the same. Say, *I started with ten fingers up. I put one finger down. I now have nine fingers up.* Write the number sentence $10 - 1 = 9$. Continue this way until you have all ten fingers down. After each, have students describe the situation and write the number sentence using the sentence frame: **I started with ten fingers up. I put ____ fingers down. I now have ____ fingers up. 10 − fingers down = fingers up.**

English Language Development Leveled Activities

Emerging Level	Expanding Level	Bridging Level
Act It Out	**Developing Oral Language**	**Think-Pair-Share**
Select ten students. Position them as a group of ten. Then, have two students from the group stand to the side. Say, *Ten minus two equals eight. The difference is eight.* Have students repeat chorally. Reposition students into a group of ten and say, *Ten minus zero equals ten. The difference is ten.* Position the ten students into other combinations, including $10 - 10 = 0$. After positioning the students, name the number sentence they are demonstrating, and have students repeat chorally	Review the word *equal* and its Spanish cognate, *iqual.* Distribute ten two color counters to each student. Have students model with you, each of the tens subtraction facts. Start with ten minus zero and end with ten minus ten. Have students take turns reciting the representation of each number sentence using this sentence frame: **____ plus ____ equals ten.**	Model pairs of combinations of ten, such as $10 - 6 = 4$ and $10 - 4 = 6$, using counters. Have students say the number sentences. Distribute two color counters to students. Have pairs work together to create the combinations of ten. If one student creates one yellow and nine red for $10 - 1 = 9$, the other student should create nine yellow and one red for $10 - 9 = 1$. Have pairs write the number sentences in their math journals. Have pairs share their work with another pair or the whole group.

Teacher Notes:

NAME _____ DATE _____

Lesson 12 Vocabulary Word Study
Subtract from 10

Circle the correct word to complete the sentence.

I. The difference is the answer to a _____ problem.

(subtraction)   addition

Show what you know about the word:

difference

There are ___10___ letters.

There are ___4___ vowels.

There are ___6___ consonants.

___10___ letters − ___4___ vowels = ___6___ consonants.

Draw a picture to show what the word means.

See students' examples.

Teacher Directions: Provide a description, explanation, or example of the new term using images or real objects. Read the sentence and have students circle the correct word. Direct students to count the letters, vowels and consonants in the math term, then complete the subtraction number sentence. Guide students to draw a picture representing their math term. Then encourage students to describe their picture to a peer.

Grade 1 • **Chapter 2** Subtraction Concepts   **27**

Lesson 13 Relate Addition and Subtraction

English Learner Instructional Strategy

Collaborative Support: Act It Out

Write the term related facts on the board. Say, *Related facts are a group of four facts that use the same three numbers.*

Have students call out two numbers that are less than six. Write the numbers on large cards. Create the addition number sentence for the sum of those numbers and write the sum on another large card. For example 2 + 3 = 5. Say, *Using these numbers (two, three, and five) I will create three more number sentences.* Rearrange the cards to create 3 + 2 = 5, 5 − 2 = 3 and 5 − 3 = 2.

As an extension after the lesson, provide pairs of students with 3 numbers that are related facts. Have them create three cards with the numbers, one with a minus sign, one with an addition sign and one with an equals sign. Students then model making related facts.

English Language Development Leveled Activities

Emerging Level	Expanding Level	Bridging Level
Act It Out Write the related facts on the board: 2 + 4 = 6, 6 − 4 = 2, 6 − 2 = 4. Select 6 volunteers. Create a group of 4 and a group of 2. Read each number sentence aloud. As you read each digit, have the appropriate group shake their hands. When the number 6 is stated, both groups should shake their hands. Repeat the activity with the different related facts. Have students read the number sentences aloud and group themselves to reflect the related facts.	**Synthesis** Write the related facts 5 + 3 = 8, 8 − 3 = 5, and 8 − 5 = 3. Explain that these are related facts. Divide students into groups of 4. Say, *One plus three equals four.* Write 1 + 3 = 4 on the board. Have one student in each group say and write the fact. Then, going clockwise, have other students say and write a related fact. Continue until all students have had an opportunity to participate.	**Think-Pair-Share** Distribute connecting cubes to pairs of students. Have one partner model an addition fact, such as 6 + 3 = 9. The other student will write and say the number sentence and then use the cubes to model a related fact, such as 9 − 6 = 3. Have pairs continue with different examples. Have students share their work with another pair or the whole group using the sentence frame: ____ − ____ = ____ and ____ + ____ = ____ are **related facts.**

Teacher Notes:

NAME _____ DATE _____

Lesson 13 Note Taking
Relate Addition to Subtraction

Read the question. Write words you need help with. Use your lesson to write your Cornell notes. Write or draw math examples to explain your thinking. Share your examples with a classmate.

Building on the Essential Question	**Notes:**
How can I relate addition to subtraction?	Related facts use the __same__ __numbers__.
	You can write related addition and subtraction __facts__.
	$2 + 7 = 9$ $9 - 7 = 2$
	$7 + 2 = 9$ $9 - 2 = 7$
Words I need help with:	These facts can help you __add__ and __subtract__.
See students' words.	You can use $2 + 7 = 9$ to find $9 - 2 = $ __7__.

My Math Examples:

See students' examples.

 Teacher Directions: Read the Building on the Essential Question and have students list words/phrases they need assistance with. Provide descriptions, explanations, or examples of the terms using images or real objects. Read each sentence frame and have students write the appropriate terms. Have students read their notes aloud. Direct students to draw a picture representing the question. Then encourage students to describe their picture to a peer.

Lesson 14 True and False Statements

English Learner Instructional Strategy

Collaborative Support: Think-Pair-Share

Review the terms true and false. Make statements such as, *I am spinning in circles or I have teeth in my mouth*. Discuss whether these statements are true or false. Explain that subtraction number sentences are also statements that can be true or false. Write some true and false subtraction number sentences and ask students to determine if they are true or false. Have students give a thumbs-up to indicate if the statement is true and a thumbs-down to indicate if the statement is false. Have students work in pairs. Assign Exercises 7–18. First have students determine individually if the statement is true or false. Then have pairs take turns sharing their answers using the sentence frame: **For number _____, the statement is (true/false)**. *What was your answer?*

English Language Development Leveled Activities

Emerging Level	Expanding Level	Bridging Level
Act It Out	**Show What You Know**	**Developing Oral Language**
Place a train of 10 connecting cubes in one side of a bucket balance. Place 10 in the other bucket to show they balance/equal. Remove one set of cubes. Say, *Three plus four equals ten. Is this true or false?* Place the 7 cubes in the other side of the balance. Say, *No, three plus four does not equal ten.* The number sentence is false. Add 3 cubes to balance the scale. Say, *Three plus seven equals ten.* Allow students to use connecting cubes to model other true and false math facts. If false, have students correct it to make it true.	Write several true and false subtraction sentences on the board. Have students come to the board one at a time. Students should choose a number sentence and read it aloud. Have each student circle the statement if it is true or put an X through it if the statement is false. If the statement is false, have students explain why it is false. Discuss with students what manipulatives they could use to determine whether a statement is true or false. Have students model their thinking.	Distribute number cards from 0 to 10 to each student in a pair. Have both students flip over a card. One student makes a subtraction number sentence and the other student identifies it as true or false. If true, the first student has to prove it is true by modeling with appropriate manipulatives. If false, the second student has to show why it is false. Continue until all cards have been used. Have pairs share their experience with the whole group.

Teacher Notes:

NAME _____ DATE _____

Lesson 14 Note Taking
True and False Statements

Read the question. Write words you need help with.
Use your lesson to write your Cornell notes. Write
or draw math examples to explain your thinking.

Building on the Essential Question	**Notes:**
How do I know if a statement is true or false?	Statements can be <u>true</u> or <u>false</u>. A <u>true</u> statement is correct. 3 − 1 = 2 A <u>false</u> statement is **in**correct or wrong. incorrect ⇩ 3 − 1 = 1 I can use <u>related</u> <u>facts</u> to see if the difference is correct. 1 + 2 = 3 3 − 1 = 2 2 + 1 = 3 3 − 2 = 1
Words I need help with: See students' words.	
My Math Examples: See students' examples.	

 Teacher Directions: Read the Building on the Essential Question and have students list words/phrases they need assistance with. Provide descriptions, explanations, or examples of the terms using images or real objects. Read each sentence frame and have students write the appropriate terms. Have students read their notes aloud. Direct students to draw a picture representing the question.

Grade 1 • **Chapter 2** *Subtraction Concepts* **29**

Chapter 3 Addition Strategies to 20

What's the Math in This Chapter?

Mathematical Practice 1: Make sense of problems and persevere in solving them

Place a messy pile of 15 books where the class can see. Hold a stack of 5 more books. Think aloud to model making sense of your problem. Say, *The reading group starts soon! Do I have enough books? I need to make sense of this problem fast.* Place the stack of 5 books next to the pile. Point to the stack and say, *I don't have time to add these to the pile and recount them all.* Point to the pile and say, *But I know there are 15 in the pile. My plan, or strategy, is to start at 15 and count on as I add the rest.* Count on from 15, adding each extra book to the pile, until you reach 20. Say, *Yes! I have 20 books.* Ask, *Have you ever made sense of an addition problem?* Solicit **Yes** from students, then have students share their experiences of making sense of a problem to create a strategy. Ask, *What strategies have you used to help you solve an addition problem?* Have students discuss with a peer then as a group. The goal is to have students recognize that making sense of a problem and creating a strategy will help them.

Display a chart with Mathematical Practice 1. Restate Mathematical Practice 1 and have students assist in rewriting it as an "I can" statement, for example: **I can make sense of my problem and make a plan to solve it.** Have students draw or write examples of when they made sense of a math problem. Post the new "I can" statement and examples in the classroom.

Inquiry of the Essential Question:

How do I use strategies to add numbers?

Inquiry Activity Target: **Students come to a conclusion that making sense of addition problems and using addition strategies to mentally figure out facts is important to solving them.**

As an introduction to the chapter, present the Essential Question to students. The inquiry graphic organizer will offer opportunities for students to observe, make inferences, and apply prior knowledge of addition strategies representing the Essential Question. As they investigate, encourage students to draw, write, and collaborate with peers to demonstrate their observations and thinking. Then have students present additional questions they may have to a peer to extend discussions.

Regroup students and restate Mathematical Practice 1 and the Essential Question. Pose questions to reflect on what has been learned to guide students in making connections between the Mathematical Practice and the Essential Question.

NAME _____ DATE _____

Chapter 3 Addition Strategies to 20
Inquiry of the Essential Question:

How do I use strategies to add numbers?

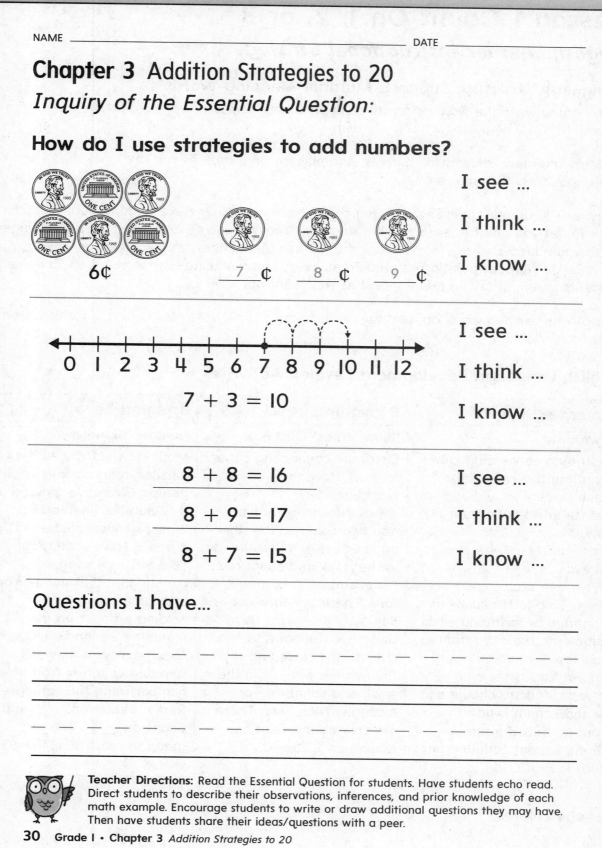

6¢   7 ¢   8 ¢   9 ¢

I see ...

I think ...

I know ...

7 + 3 = 10

I see ...

I think ...

I know ...

8 + 8 = 16

8 + 9 = 17

8 + 7 = 15

I see ...

I think ...

I know ...

Questions I have...

_____

_ _ _ _ _ _ _ _ _ _ _ _ _ _ _ _ _ _

_____

_ _ _ _ _ _ _ _ _ _ _ _ _ _ _ _ _ _

_____

Teacher Directions: Read the Essential Question for students. Have students echo read. Direct students to describe their observations, inferences, and prior knowledge of each math example. Encourage students to write or draw additional questions they may have. Then have students share their ideas/questions with a peer.

Lesson 1 Count On 1, 2, or 3
English Learner Instructional Strategy

Language Structure Support: Multiple-Meaning Words

Say, *on* and write the word *on* for all students to see. Say, *The word on has multiple meanings. Let's learn about some of them.*

Pick up an object, show it to students, and place it on a desk as you say *I put the object on the desk.* Stress the word *on.*

Draw a 0–10 number line. Say, *Let's count to the number five.* Have students count aloud with you as you point to each number and say, *One, two, three, four, five.* Keep your finger by the number 5 and say, *I will count on.* Point to the numbers as you count on, saying, *Six, seven, eight.* Repeat pointing at random numbers and modeling how to count on. Encourage students to join in saying the numbers as you count on.

Point to the written word, on, and say, *When you count on you are counting forward from a certain number.*

English Language Development Leveled Activities

Emerging Level	Expanding Level	Bridging Level
Synthesis	**Show What You Know**	**Number Game**
Distribute connecting cubes. Create a five-cube train. Each student should also create a five-cube train. Say, *Now, I will add two more cubes. I know this is five cubes. I will start with five. I will* **count on** *two more.* Emphasize *count on.* Continue by snapping each cube onto the cube train as you say, *Six cubes, seven cubes.* Ask, *Is the sum seven?* Students should nod or show thumbs-up to confirm that the sum is seven. Repeat counting on with 1, 2 or 3.	Distribute connecting cubes to pairs. Have them stack two cubes. Say, *Add two cubes.* Have one student add two cubes to the stack and say the number already in the stack and count on. For example, **Two, three, four. There are four cubes.** Say, *Add one cube.* Have the other student add one cube to the stack and say the number already in the stack and count on. For example, **Four, five. There are five cubes.** Continue adding on 1, 2, or 3.	Model this activity with a partner, then work in small groups. Give each group a 0–12 number line and a 3-part spinner numbered 1, 2 and 3. Have each group take turns spinning, counting on that number on the number line, and holding a finger on the number they land on. On their next turn, they will spin and count on from that number using this sentence frame: **I start on _____ and count on _____ .** The first group to reach 12 is the winner.

Teacher Notes:

NAME _____

DATE _____

Lesson 1 Note Taking
Count On 1, 2, or 3

Read the question. Write words you need help with. Use your lesson to write your Cornell notes. Write or draw math examples to explain your thinking.

Building on the Essential Question	**Notes:**
How can I count on to add 1, 2, or 3?	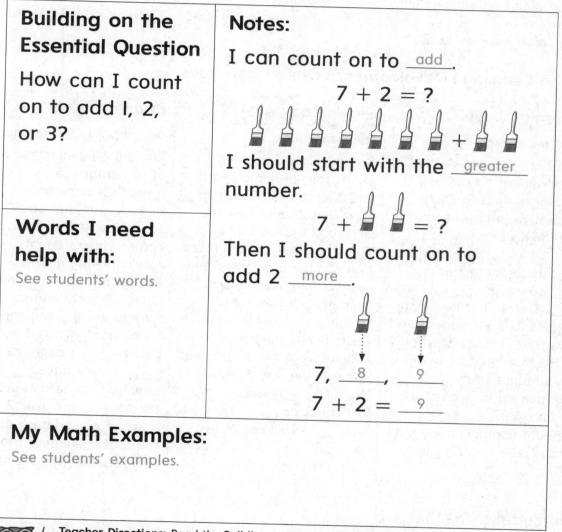
Words I need help with: See students' words.	

My Math Examples: See students' examples.

Teacher Directions: Read the Building on the Essential Question and have students list words/phrases they need assistance with. Provide descriptions, explanations, or examples of the terms using images or real objects. Read each sentence frame and have students write the appropriate terms and numbers. Have students read their notes aloud. Direct students to draw a new picture representing the question. Then encourage students to describe their picture to a peer.

Grade 1 · **Chapter 3** Addition Strategies to 20   **31**

Lesson 2 Count On Using Pennies

English Learner Instructional Strategy

Collaborative Support: Partners Work

Have students work in pairs during the Explore and Explain activity. Model how to *count on* using pennies with a volunteer. Count 3 pennies then have the student *count on* 3 more. Distribute 12 pennies to each pair of students. Say, *Reese has six pennies in his bank.* **Have** one student count out 6 pennies. Then say, *He puts in three more pennies.* **Have the other** student count out 3 pennies. Ask, *How many pennies does Reese have in all? We will count on to find out.* Have the student with 6 pennies count each penny out loud. Have the other student count on, starting with 7. *Ask, How many pennies in all ?* **nine** Write the number sentence, 6 + 3 = 9. Say, *Six pennies and three pennies is nine pennies in all.* Elicit students to say, **nine pennies in all.**

English Language Development Leveled Activities

Emerging Level	Expanding Level	Bridging Level
Number Sense Distribute two pennies to each student. Say, *Count your pennies.* Give each student two more pennies. Then say, *Start with two and count on.* Have students use this sentence frame: _____ **count on** _____. **I have** _____. Distribute three more pennies to each student and have them repeat, starting on the last number, using the sentence frame. Continue up to twelve pennies. Repeat with different numbers of pennies to start and counting on with 1, 2, or 3.	**Number Game** Distribute pennies and a 3-part spinner numbered 1, 2, and 3 to pairs of students. Have one student spin and say aloud the number spun. Then the other student takes that number of pennies, and counts them aloud. Have each partner take turns spinning and counting out the required number of pennies. Together pairs count on using this sentence frame: **We had** _____ **pennies. We will count on** _____. **Now we have** _____ **pennies.** When pairs count up to 12, have them start over.	**Developing Oral Language** Distribute pennies and a set of 10 number cards (randomly numbered 1, 2, or 3) to small groups. Cards should be face down in the center. Have students take turns drawing a card and taking that many pennies. On subsequent turns, players use this sentence frame as they count on: **I had** _____. **I count on** _____. **Now I have** _____ **pennies.** If a card causes someone to go over 12, it must be placed back and the turn is skipped. The first to reach exactly 12 wins.

Multicultural Teacher Tip

Most ELs have had math education in their native countries and are familiar with basic math concepts. However, mathematical discourse in an unfamiliar language can be intimidating and confusing, and students may struggle with even seemingly simple steps leading to a solution. Manipulatives are a helpful option for EL students. By utilizing concrete objects to model familiar concepts or to learn new ones, students can work around language barriers that might make verbal or written explanations too difficult. Keep in mind that manipulatives are not always used in other cultures, and the student may need time and encouragement to become comfortable using them.

NAME _____ DATE _____

Lesson 2 Four-Square Vocabulary
Count On Using Pennies

Trace the word. Write the definition for *count on*.
Write what the words mean, draw a picture, and
write your own sentence using the words.

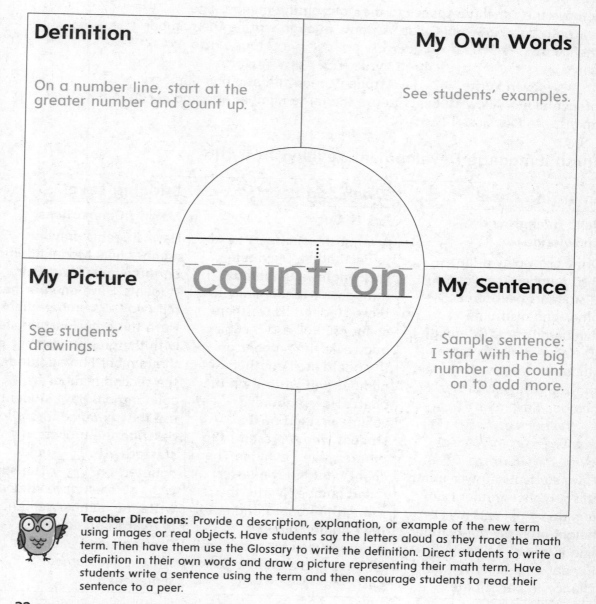

Definition

On a number line, start at the
greater number and count up.

My Own Words

See students' examples.

My Picture

See students'
drawings.

count on

My Sentence

Sample sentence:
I start with the big
number and count
on to add more.

Teacher Directions: Provide a description, explanation, or example of the new term
using images or real objects. Have students say the letters aloud as they trace the math
term. Then have them use the Glossary to write the definition. Direct students to write a
definition in their own words and draw a picture representing their math term. Have
students write a sentence using the term and then encourage students to read their
sentence to a peer.

Lesson 3 Using a Number Line to Add

English Learner Instructional Strategy

Graphic Support: Number Line

Draw or display a large 0–12 number line. The number line should span across the front of the room as widely as possible. Draw a bridge over the number line, similar to the Explore and Explain activity. Have a large image of a car ready for the Use a Number Line to Add Activity. Hold up the image of the car. Say, *A car is driving across the bridge. It starts at the number four.* Place the car on the number 4 on the number line. *It drives three spaces to the right.* Have students count, one, two, three, as you move the car three spaces to the right on the number line. *Where does the car stop?* **Seven** Guide students through writing the number sentence. Ask, *On which number did the car start?* **four** Write 4. Ask, *How many spaces did it drive?* **three** Write + 3. Ask, *On which number did the car stop?* **seven** Write = 7. Have students read the addition number sentence aloud. Allow students to use the large number line and car to demonstrate Exercises 1–19.

English Language Development Leveled Activities

Emerging Level	Expanding Level	Bridging Level
Build Background Knowledge Draw or display a large, 0–10 number line. Position 11 students next to each other and distribute numbered cards 0 through 10, sequentially to these students. Have each student show and say his or her number. Say, *Your neighbor on the number line is the number next to you.* Ask, *Who are your neighbors?* Have students answer using this sentence frame: **I am number ____ and my neighbor(s) on the number line is/are ____ and ____.** Repeat until all have had a chance to participate.	**Act It Out** Arrange 11 volunteers next to each other. Distribute sticky notes, numbered 0 through 10, sequentially to these students. Have them count off. Roll a 0–5 cube and write the number on the board. Roll another number and write it on the board. Have another volunteer stand by the student who represents the greater rolled number. Then count on using the lesser rolled number. Write the sum on the board and read the addition number sentence aloud. Repeat until all have had a chance to participate.	**Making Connections** Have students draw a 0–12 number line. Model addition on the number line. Pair students. Have one partner roll two 0–5 number cubes. Have the other partner start with the greater number on the number line and add the second number by counting on. Have students use this sentence frame to describe the process: I **started with ____ and counted on ____. The sum is ____.** Then have students write the number sentence in their Math Journals. Continue until each pair has written five number sentences.

Teacher Notes:

NAME _____ DATE _____

Lesson 3 Vocabulary Definition Map
Use a Number Line to Add

Trace the term. Use the definition map to write what the math term means and tell what the term is like. Write or draw a math example. Share your examples with a classmate.

My Math Word:

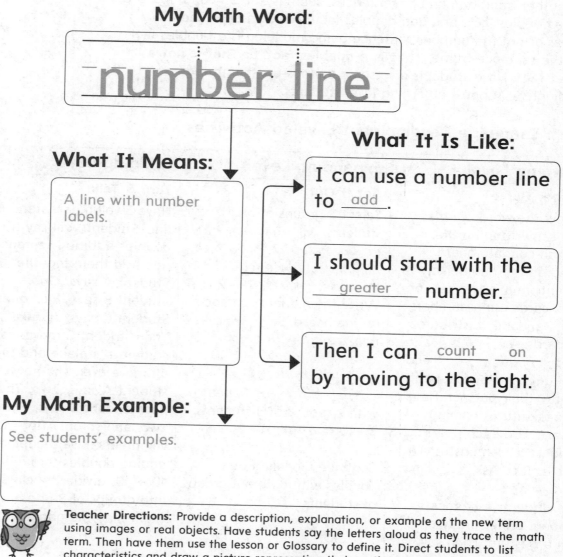

number line

What It Means:

A line with number labels.

What It Is Like:

I can use a number line to __add__.

I should start with the __greater__ number.

Then I can __count__ __on__ by moving to the right.

My Math Example:

See students' examples.

Teacher Directions: Provide a description, explanation, or example of the new term using images or real objects. Have students say the letters aloud as they trace the math term. Then have them use the lesson or Glossary to define it. Direct students to list characteristics and draw a picture representing their math term. Then encourage students to describe their picture to a peer.

Grade I • **Chapter 3** *Addition Strategies to 20* **33**

Lesson 4 Use Doubles to Add

English Learner Instructional Strategy

Vocabulary Support: Cognates

Write the word *doubles* and the Spanish cognate, *dobles* on a cognate chart. Say, *When both **addends** in a number sentence are the same number, it is a **doubles fact***. Addends are the numbers being added together.

Write the number sentence 2 + 2 = 4. Say, *Two plus two equals four. This is a doubles fact because both addends are two.*

Model other addition number sentences, such as 3 + 3 = 6; 5 + 3 = 8. Include doubles facts and non-doubles facts. Have students identify whether or not the number sentence being modeled is a doubles fact using the sentence frame: **That is a doubles fact.** or **That is not a doubles fact.** Have students also show thumbs-up/thumbs-down accordingly to support emerging level students.

English Language Development Leveled Activities

Emerging Level	Expanding Level	Bridging Level
Number Game	**Act It Out**	**Turn & Talk**
Review the word *doubles* (*dobles*) written on the Spanish cognate chart. Based on the number of students, make two sets of index cards using numbers 0–10. Randomly distribute the cards. Ask, *Who has the same number as you? Find your double.* Have students identify their partner. Have pairs take turns coming forward. Use each pair's numbers to demonstrate a doubles fact. Ask, *What is the sum of these doubles?* Give students a chance to respond, either orally or with a gesture. Then say, *The sum is 14.*	Select 2 groups of 2 students. *Ask, How many are in your group?* **2** *What is the number sentence to add these two groups?* **2 + 2 = 4** Write the number sentence on the board. Say, *Two and two are addends in this number sentence. When addends are the same, they are called doubles.* Repeat with 2 groups of 3 students. **3 + 3 = 6** Ask, *Are three and three doubles?* **Yes** Continue modeling other doubles with different groups of students.	Have 3 volunteers stand in a line. Students will say a doubles addition sentence and add them together. Student A says, **One.** Student B says, **Plus one.** Student C says, **Is two!** Then student C moves to student A's place and they all move over. The new student A says, **Two.** The new student B says, **Plus two**, and so on. After the volunteers have modeled adding doubles through 10 + 10, divide the class into groups of 3 and have them play the game.

Teacher Notes:

NAME _____ DATE _____

Lesson 4 Vocabulary Sentence Frames
Use Doubles to Add

The math words in the word bank are for the sentences below. Write the words that fit in each sentence on the blank lines.

Word Bank		
doubles	addends	sum

1. The answer to an addition problem is the ___sum___.

   4 + 5 = 9
           ↑

2. ___Addends___ are the numbers you add.

   4 + 5 = 9
   ↑   ↑

3. Both addends are the same in a ___doubles___ fact.

   (4 + 4) = 8

Teacher Directions: Provide a description, explanation, or example of the each term using images or real objects. Read each sentence frame and have students echo read. Direct students to write the correct terms in each blank. Then encourage students to read each sentence to a peer.

34   Grade 1 • Chapter 3 *Addition Strategies to 20*

Lesson 5 Use Near Doubles to Add

English Learner Instructional Strategy

Collaborative Support: Think-Pair-Share

Before Explore and Explain, model adding the doubles fact, $3 + 3 = 6$, with connecting cubes. Ask, *What is the sum?* **six** Write $3 + 3 = 6$.

Display two groups of 3 connecting cubes. Say, *Here are two groups of three. I will remove one cube.* Remove 1 connecting cube from a group of 3, to make 2. Hold up the group of 2 and ask, *How does this group differ from the other?* **It has one less cube.** Hold up both groups and say, *I removed one cube from the group, so together they have one less than three and three. The sum of 3 + 2 will be one less than the sum of 3 + 3.* Write $3 + 2 = ?$. Ask, *What is one less than the sum of three plus three?* **five** Complete the number sentence by writing the sum, 5. Repeat modeling adding **one more** to $3 + 3 = 6$ to have $3 + 4 = 7$.

Have students work in pairs using near doubles to solve Exercises 3–17. Have pairs share their findings with another pair or the whole group.

English Language Development Leveled Activities

Emerging Level	Expanding Level	Bridging Level
Number Sense	**Number Recognition**	**Making Connections**
Distribute one color of connecting cubes. Write $2 + 2 = 4$. Have students model the sentence and say, **Two plus two equals four.** Ask, *Are the addends doubles?* **yes** Students should nod or show thumbs-up. Give each student one different color connecting cube. Have students place the different color cube on one of their trains. Ask, *What is two plus two and one more?* Encourage students to say with you, **Two plus two is four. Four plus one is five.** Repeat with other sets of doubles.	Distribute 10 counters to each pair of students. Have one student place 3 color counters in each hand. Have student B write and say the number sentence: $3 + 3 = 6$. Say, *doubles plus one,* and distribute one different color counter to each pair. Have student B place the counter in one of student A's hands. Have student B write "$+ 1 = 7$" after the number sentence and say: $3 + 3 + 1 = 7$. Have partners switch roles and repeat the activity with different doubles facts minus 1.	Provide pictures or show realia (concrete objects) of different sets of doubles. For example, show two packs of five pencils. Have students suggest other items that come in multipacks of two, three, four, five, or six that can represent doubles. For each example, ask, *How many would you have if you had one more/one less?* Have students find the sum of the doubles and then add or subtract one and explain using the words **doubles plus one** or **doubles minus one.**

Teacher Notes:

NAME _____ DATE _____

Lesson 5 Vocabulary Identification
Use Near Doubles to Add

Match each term to a picture.

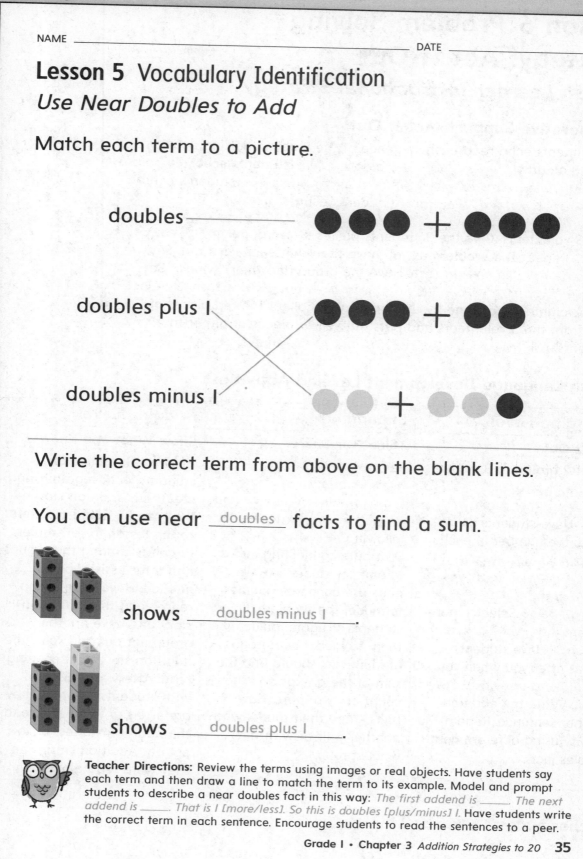

doubles

doubles plus 1

doubles minus 1

Write the correct term from above on the blank lines.

You can use near __doubles__ facts to find a sum.

shows ___doubles minus 1___.

shows ___doubles plus 1___.

Teacher Directions: Review the terms using images or real objects. Have students say each term and then draw a line to match the term to its example. Model and prompt students to describe a near doubles fact in this way: *The first addend is _____. The next addend is _____. That is 1 [more/less]. So this is doubles [plus/minus] 1.* Have students write the correct term in each sentence. Encourage students to read the sentences to a peer.

Grade 1 • **Chapter 3** *Addition Strategies to 20* **35**

Lesson 6 Problem Solving Strategy: Act It Out

English Learner Instructional Strategy

Collaborative Support: Act It Out

Have students echo read each sentence of the Problem Solving activity as you read aloud: *Three red birds are on a branch.* (students echo) *There are two more yellow birds than red birds on another branch.* (students echo) *How many yellow birds there are?* (students echo)

Select 3 volunteers. Give the 3 students pieces of red paper to use as "wings." Repeat the problem as you have the volunteers act it out. Say, *Three red birds are on a branch.* Have the group flap their "wings." Say, *There are two more yellow birds than red birds on another branch.* Select 2 more volunteers to stand by the group of 3. Collect the red pieces of paper from the 3 volunteers and pass out yellow pieces of paper to all 5 students. Ask, *How many yellow birds are there?* **five**

English Language Development Leveled Activities

Emerging Level	Expanding Level	Bridging Level
Act It Out	**Think-Pair-Share**	**Show What You Know**
Select 2 groups of 3 students. Say, *Here is a group of three. Here is another group of three.* Have students count aloud, the number in each group. Ask, *Is three plus three a* **doubles fact?** **yes** *How many students are there in all?* **6** Select 1 more student. Ask, *Now, how many students?* Have students repeat after you when you say, *Six plus one equals seven.* Write the addition number sentence. Repeat the activity using different near doubles facts.	Say, *Four students play. Four more students play. If two more join, how many will be playing in all?* Write the following: $4 + 4 = ? + 2 = ?$ Distribute connecting cubes to pairs of students. Have pairs use connecting cubes to model the problem. If students struggle, guide them to model two groups of four. Pairs should find the sum of the groups and then count on two more. Have pairs share their model with another pair or with the whole group.	Have small groups of students work together and create an addition story that can be acted out. Refer to the lesson for examples. Circulate among the groups and offer assistance as needed. Have each group read aloud their addition story and have another group act out and solve the addition story as it is being read. Afterward, discuss with students whether they preferred to write and read the addition stories, or to act out addition stories as they were read aloud.

Teacher Notes:

NAME _____ DATE _____

Lesson 6 Problem Solving
STRATEGY: Act It Out

<u>Underline</u> what you know. (Circle) what you need to find. Act out the problem to find the answer.

I. <u>A hot dog **vendor** sold **9** hot dogs on **Monday**.</u>

 hot dog

<u>**She** (the vender) sold the **same number** of hot dogs on **Tuesday**.</u>

How many hot dogs did **she** sell **in all**?

 vendor

Act it out

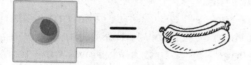

__ 9 (+) __ 9 (=) __ 18

She sold __18__ hot dogs in all.

 Teacher Directions: Provide a description, explanation, or example of the boldface terms and nouns using images or real objects. Read each sentence and have students echo read. Have students use connecting cubes to act out the problem. Then encourage them to use the part-part-whole mat to keep track of their addends and the sum. Direct them to write their answer in the restated question. Have students read the answer sentence aloud.

Lesson 7 Make 10 to Add

English Learner Instructional Strategy

Sensory Support: Act It Out

Review numbers that make 10 before the Explore and Explain activity. Show ten fingers to students. Ask, *How many fingers?* **10** Have students show you ten fingers. Put your pinky finger down on your right hand and have students mirror your actions. Elicit students to say, *1 plus 9 equals 10.* Put the same pinky finger down along with the ring finger, have students mirror and say, *2 plus 8 equals 10.* Discuss that the 2 fingers down shows the first addend and the 8 fingers up shows the second addend. Continue acting out all numbers that make ten through 9 plus 1 equals 10.

English Language Development Leveled Activities

Emerging Level	Expanding Level	Bridging Level
Word Knowledge Show a number line. Ask students, *What is this?* Have them answer using this sentence frame: **That is a ____.** Draw or show examples of addends and sets of doubles. Ask, *What are these?* Have students answer using this sentence frame: **Those are ____ and ____.** Demonstrate counting on by starting with ten connecting cubes and adding two. Ask, *What was I doing?* **counting on.** Repeat questions with different examples of addends, counting on, doubles, and number lines to review chapter vocabulary.	**Making Connections** Gather counters and Work Mat 2. Say and write, *Find 7 + 4.* Model by filling the top ten-frame with 7 red counters and the bottom ten-frame with 4 yellow counters and lead students in a choral count to eleven. Then, model by filling the top ten-frame with 7 red counters and the bottom ten-frame with 4 yellow counters. Say, *I will move three yellow counters to make ten.* Write 10 + 1 = 11. Say, *Making a ten makes mental addition easier.*	**Number Fun** Distribute counters, 2 ten-frames, and write-on/wipe-off boards to small groups. Write 8 + 4 = on the board. Say, *Copy this addition sentence. Next to it, write 10 + = ____. Use counters and ten-frames to solve. Then complete the addition sentence.* The group that solves the number sentence first and shows how to make ten correctly wins. The winning group creates and writes the next addition sentence for other groups to solve. Continue until each group has written an addition sentence.

Teacher Notes:

NAME _____ DATE _____

Lesson 7 Sum Identification
Make 10 to Add

Match addition examples to show how to make a 10 to add.

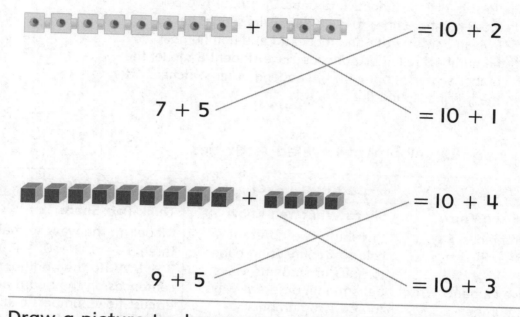

Draw a picture to show one way to make a 10 to add.

See students' examples.

Teacher Directions: Model an addition sentence, such as 8 + 3, and use manipulatives to show how to make a 10 to find the sum. Have students match an example on the left with the make-a-10 example on the right. Then direct students to draw a picture representing a way to make a 10 when adding. Finally encourage students to describe their picture to a peer.

Grade 1 • Chapter 3 *Addition Strategies to 20* **37**

Lesson 8 Add in Any Order
English Learner Instructional Strategy

Sensory Support: Hands On Activities

Before Explore and Explain, select two volunteers. Have each volunteer model an addition sentence using counters. Have one volunteer model 4 + 3 and the other volunteer model 3 + 4. Ask, *What is the sum of four plus three?* **seven** *What is the sum of three plus four?* **seven** *How are the addition number sentences different from each other?* **The addends are switched.** Say, *The sum is the same no matter the order of the addends.* As an extension to the lesson, divide students into pairs. Distribute two-color counters and two 0–5 number cubes to each pair. Pairs should roll the number cubes. Have student A write and model an addition number sentence using the numbers rolled as addends. Have student B model the same addition number sentence but with the addend order switched. After each student in the pair has found the sum, ask, *Are your sums the same?* **yes**

English Language Development Leveled Activities

Emerging Level	Expanding Level	Bridging Level
Show What You Know Distribute dominoes. Say, *Write a sentence that adds the numbers on your domino.* Have students write an addition number sentence, for example, 2 + 3 = 5. Then have students rotate the domino and write another addition number sentence, 3 + 2 = 5. Circulate among the students and offer assistance as needed. Ask questions such as: *Which numbers are the addends?* **2 and 3** and *Does the order of the addends change the sum?* **no** Collect dominoes, redistribute, and repeat the activity.	**Share What You Know** Distribute two different colors of connecting cubes to pairs of students. Have pairs model different ways to add two numbers together. For example, say, *Show me two ways to add the numbers one and four.* Pairs should model the addition number sentences: 1 + 4 = 5; 4 + 1 = 5. Discuss how the order of the addends can be changed and we still get the same sum. Repeat using different number combinations with sums up to 20.	**Think-Pair-Share** Model all the ways to make three: 0 + 3, 3 + 0, 1 + 2, 2 + 1. Write the addition sentences on the board and discuss how numbers can be added in any order to get the same sum. Have students use manipulatives to model all the ways to make 5. Then have students write the addition numbers sentences in their Math Journals, pairing up ones that use the same addends. Repeat the activity with students working in pairs and a new sum such as 12.

Teacher Notes:

NAME _____ DATE _____

Lesson 8 Concept Web
Add in Any Order

Use numbers or write a word from the word bank to complete each sentence in the concept web.

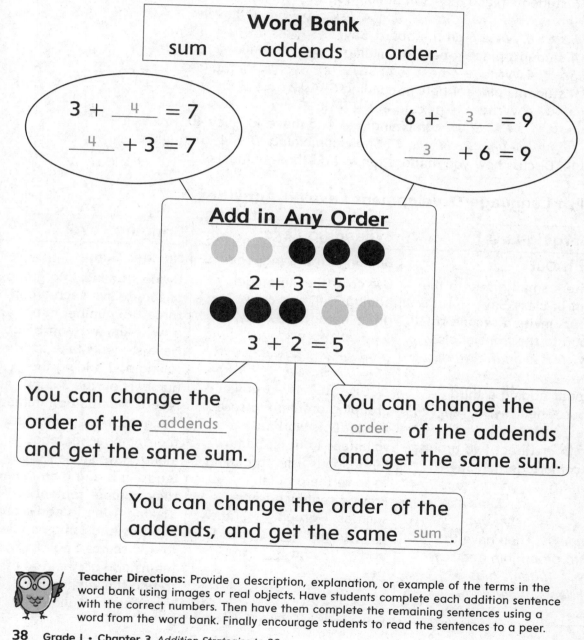

Word Bank

sum   addends   order

$3 + \underline{\ 4\ } = 7$

$\underline{\ 4\ } + 3 = 7$

$6 + \underline{\ 3\ } = 9$

$\underline{\ 3\ } + 6 = 9$

Add in Any Order

$2 + 3 = 5$

$3 + 2 = 5$

You can change the order of the _addends_ and get the same sum.

You can change the _order_ of the addends and get the same sum.

You can change the order of the addends, and get the same _sum_.

Teacher Directions: Provide a description, explanation, or example of the terms in the word bank using images or real objects. Have students complete each addition sentence with the correct numbers. Then have them complete the remaining sentences using a word from the word bank. Finally encourage students to read the sentences to a peer.

Lesson 9 Add Three Numbers

English Learner Instructional Strategy

Sensory Support: Illustrations/Pictures

Create a scene using construction paper or colored chalk/markers that includes three buildings labeled A, B, and C similar to the Explore and Explain lesson. Create 14 pets, on separate sticky notes and distribute to 14 students.

Have 6 students place 6 pets in building A. Gesture to the scene and say, *There are six pets living in building A.* **Ask,** *How many pets are in building A?* **6** Write 6 on the board. Say, *Four pets live in building B.* Have 4 students place 4 pets in building B. Ask, *How many pets are in building B?* **4** Write + 4 next to 6. Say, *Four pets live in building C.* Have 4 students place 4 pets in building C. Ask, *How many pets are in building C?* **4** Write + 4 next to 6 + 4. Ask, *How many pets are in buildings A and B?* Show combining 6 + 4 to make 10. Elicit **10**. *How many pets are in buildings A, B, and C?* Show combining 10 + 4 to equal 14. Elicit **14**. Discuss how you made a ten to add three numbers.

English Language Development Leveled Activities

Emerging Level	Expanding Level	Bridging Level
Act It Out	**Building Oral Language**	**Number Game**
Invite a small group to the front of class. Say, *This is a group.* Invite a second small group to the front of class. Say, *This is another group. There are two groups.* Repeat adding a third group. Then say, *Now I will group all the students together.* Direct two groups to come together. Say, *I added two groups together first. ____ students plus ____ students equals ____ students.* Then have the third group join and say, *____ students plus ____ students equals ____ students.*	Use color tiles to model adding 3, 4, and 6 by grouping addends. Say, *First I group two addends. I will group 4 and 6 to make 10. Then I add 10 and 3 to get 13.* Give students a similar grouping problem to solve. As students work, ask a volunteer to describe how he or she is using grouping to solve the problem. Provide sentence frames, such as: **I group ____ and ____ to get ____. Then I add ____ and ____ to get ____.**	Divide students into groups of three. Give each group three 0–5 number cubes and a write-on/wipe-off board. Have students take turns rolling and adding the numbers on the 3 cubes and record the addition number sentence. Have groups say their number sentence aloud. Have students play ten rounds and then review their number sentences. Making a ten to add scores 2 points and using doubles to add scores 1 point. Have teams add up their points to find the winner. Play again as time allows.

Teacher Notes:

NAME _____ DATE _____

Lesson 9 Note Taking
Add Three Numbers

Read the question. Write words you need help with. Use your lesson to write your Cornell notes. Write or draw math examples to explain your thinking.

Building on the Essential Question	Notes:
How can I use strategies to add three numbers?	I can group <u>numbers</u>. I can add in any <u>order</u>. I can look for <u>doubles</u>. I can make a <u>10</u>. Then I can add the other number to find the <u>sum</u>.

Words I need help with:
See students' words.

My Math Examples:
See students' examples.

 Teacher Directions: Read the Building on the Essential Question and have students list words/phrases they need assistance with. Provide descriptions, explanations, or examples of the terms. Read each sentence frame and have students write the appropriate terms. Have students read their notes aloud. Direct students to draw a picture representing the question. Then encourage students to describe their picture to a peer.

Grade 1 • **Chapter 3** *Addition Strategies to 20* **39**

Chapter 4 Subtraction Strategies to 20

What's the Math in This Chapter?

Mathematical Practice 1: Make sense of problems and persevere in solving them

Draw a pyramid of 9 squares on the board. Show a pile of 17 blocks. Say, *I need to build 2 pyramids like the one on the board. I have 17 blocks total and each pyramid takes 9 blocks.* Act confused. Write $17 - 9 = $ _____ on the board. Say, I don't remember this subtraction fact! I must persevere to find a different plan, or strategy. Model subtraction by making 10 using the blocks and writing it on the board. Then build 1 complete and 1 incomplete pyramid. Say, *I solved my problem. I didn't know the subtraction fact. I persevered and used the making 10 strategy to see I'm 1 block short.* Ask, *Have you ever tried more than 1 strategy, or plan, to solve a problem?* Solicit **Yes** from students, then have students share their experiences with perseverance. Ask, *What strategies have you used to help you solve a subtraction problem?* Have students discuss with a peer then as a group. The goal is to recognize that they can persevere to solve a problem by using different strategies. Display a chart with Mathematical Practice 1. Restate Mathematical Practice 1 and have students assist in rewriting it as an "I can" statement, for example: **I can make a plan, carry out my plan, and solve the problem.** Have students draw or write examples of when they made sense of a math problem. Post the new "I can" statement and examples in the classroom.

Inquiry of the Essential Question:

What strategies can I use to subtract?

Inquiry Activity Target: **Students come to a conclusion that subtraction strategies are key to persevering in solve a problem.**

As an introduction to the chapter, present the Essential Question to students. The inquiry graphic organizer will offer opportunities for students to observe, make inferences, and apply prior knowledge of subtraction strategies representing the Essential Question. As they investigate, encourage students to draw, write, and collaborate with peers to demonstrate their observations and thinking. Then have students present additional questions they may have to a peer to extend discussions.

Regroup students and restate Mathematical Practice 1 and the Essential Question. Pose questions to reflect on what has been learned to guide students in making connections between the Mathematical Practice and the Essential Question.

NAME _____ DATE _____

Chapter 4 Subtraction Strategies to 20
Inquiry of the Essential Question:

What strategies can I use to subtract?

$7 - 3 = 4$

I see ...

I think ...

I know ...

$5 + 5 = 10$

$10 - 5 = 5$

I see ...

I think ...

I know ...

12
5   7

$5 + 7 = 12$ $12 - 5 = 7$

$7 + 5 = 12$ $12 - 7 = 5$

I see ...

I think ...

I know ...

Questions I have...

_____

_____

_____

_____

Teacher Directions: Read the Essential Question for students. Have students echo read. Direct students to describe their observations, inferences, and prior knowledge of each math example. Encourage students to write or draw additional questions they may have. Then have students share their ideas/questions with a peer.

Lesson 1 Count Back 1, 2, or 3

English Learner Instructional Strategy

Language Structure Support: Multiple Word Meanings

Write the word *back* on the board. Say, *The word **back** has multiple meanings.* Stand with your back to students. Say *This is my **back**.* Display a book, point to the back and say, *This is the **back** of the book.*

Write a 0–10 number line. *Say, Let's start on the number 5.* Have students help you identify the number 5 on the number line. Keep your finger by the number 5 and say, *I am pointing to 5 and will count **back** by 3.* Point to each number as you count back, saying, *one, two, three.* Keep your finger by the number 2 and ask, *Now what number am I pointing to?* **2** Repeat modeling how to count back, having student volunteers point to numbers and chorally count back. Point to the written word, back and say, *When you **count back** you are subtracting, or taking away.*

English Language Development Leveled Activities

Emerging Level	Expanding Level	Bridging Level
Number Recognition Sing "10 Little Indians," substituting "First Graders" for "Indians." Have students create a circle and distribute cards numbered 1 through 10 randomly to 10 students. As you sing, *1 little, 2 little, 3 little first graders...* have students step forward as you sing their number. Sing, *10 little first-grade boys and girls*, when you reach 10. Continue the song, counting backward, directing students to step back into the circle as their corresponding number is sung. End with, *One little first grade boy or girl.* Repeat until all students have participated.	**Act It Out** Have students create a circle. Say, *We can count back to subtract.* Write 3 − 1 = _____. Ask 3 students to form a line in the center. Say, *We can count back to find the difference.* Point to the last student in line. Say, *I will start at three and count back one. Three, two. The difference is two.* Ask a student to write the difference in the subtraction sentence. Use this routine to practice counting back by 2 and 1.	**Exploring Language Structure** Write *back* on chart paper. Brainstorm with students different sentences or phrases with the word *back* and write them on the chart: *back to school, count back, back up, **back on track, back off, got your back, back down**, back and forth, behind your back, back out, turn your back on, **backfire on**, backpack, back talk, backyard, backward.* Ask student volunteers to pantomime each usage of *back.* Offer hints as needed. After the last pantomime, discuss the figurative meaning of the boldface idioms

Multicultural Teacher Tip

Word problems are an important part of the math curriculum, but they can be particularly challenging for ELs, and not just because of language issues. Allow students to share examples from their own cultures, including popular national sports or physical activities they participated in, foods and drinks from their culture, traditional clothing worn in their home countries, and so on. When appropriate, help ELs reword an exercise to include a familiar cultural reference.

NAME _____ DATE _____

Lesson I Word Web

Count Back I, 2, or 3

Trace the math words. Draw a number story in each rectangle that shows the meaning of *count back*.

See students' examples.

count back

See students' examples.

Teacher Directions: Provide a description, explanation, or example of the new term using images or real objects. Have students say the letters aloud as they trace the math term. Direct students to draw two picture stories that represent the math term. Have them complete the sentence. Then encourage students to tell their number stories to a peer.

Grade I • **Chapter 4** *Subtraction Strategies to 20* **4I**

Lesson 2 Use a Number Line to Subtract
English Learner Instructional Strategy

Sensory Support: Act It Out

Draw or display a large 0–12 number line. The number line should span across the front of the room as widely as possible. Have 11 large images of beach balls ready for this act it out activity. Say, *A store has eleven beach balls to sell.* Place the beach ball images above the numbers 1 through 11 on the number line. Say, *It sells two of them.* Have students count back from 11, **ten, nine** as you remove the beach balls on the numbers 11 and 10 on the number line. *How many beach balls do they have left?* **nine** Have students model using the paper clip on their number lines.

Guide students through writing the number sentence. Ask, *How many beach balls did the store **have** to sell?* **eleven** Point to 11. Ask, *How many beach balls **did** the store sell?* **two** Point to 2. Ask, *How many beach balls **does** the store have left?* **nine** Write 9. Have students read the subtraction number sentence aloud.

English Language Development Leveled Activities

Emerging Level	Expanding Level	Bridging Level
Act It Out	**Sentence Frames**	**Number Game**
Write the words *difference* and *number line* and their Spanish cognates, *diferencia* and *linea numerica* on a cognate chart. Create a 0–10 number line using a long piece of masking tape and students holding pieces of paper numbered 0 through 10. Say, *The number line shows zero through ten.* Write 8 − 2 = ____. Stand at 8. Take 2 large steps over to 6 as you, say, *eight, seven, six. The difference is six.* Repeat with exercises from the lesson using student volunteers to model counting back.	Place papers numbered 1–10 on the floor. Stand on 10 and step back two spaces, counting back as you go. Say, *I started on ten. I counted back two. Now I am on eight. The difference is eight.* Have a volunteer stand on 3. Say, *Count back 2.* Have the student use this sentence frame after moving back two spaces: **I started on ____. I counted back ____. Now I am on ____. The difference is ____.** Repeat the activity with different students and numbers.	Distribute a 0–20 number line and a game piece to each student. Place students into groups of 3 and distribute a 3-part spinner numbered 1 through 3. All students place their game piece on 20 on the number line. Students take turns spinning a number. All group members count back that number on their number lines. With each subsequent turn, students continue counting back the number spun using the sentence frame: **We start on ____ and count back ____. We land on ____. The difference is ____.** The first group to reach 0 wins.

Teacher Notes:

NAME _____ DATE _____

Lesson 2 Note Taking
Use a Number Line to Subtract

Read the question. Write words you need help with. Use your lesson to write your Cornell notes.

Building on the Essential Question	**Notes:**
How can I use a number line to subtract?	**Word Bank** count back difference greater number subtract $7 - 3 = \underline{\ 4\ }$ I can use a number line to help me __subtract__. I should start with the __greater__ __number__. 0 1 2 3 4 5 6 7 8 9 10 11 12

Words I need help with:

See students' words.

Then I can __count__ __back__ by moving to the left.

0 1 2 3 4 5 6 7 8 9 10 11 12

When I subtract, I can count back to find the __difference__.

0 1 2 3 4 5 6 7 8 9 10 11 12

Teacher Directions: Read the Building on the Essential Question and have students list words/phrases they need assistance with. Provide descriptions, explanations, or examples of the terms using images or real objects. Read each sentence frame and have students write the appropriate terms. Have students read their notes aloud.

Lesson 3 Use Doubles to Subtract

English Learner Instructional Strategy

Vocabulary Support: Activate Prior Knowledge

Review the word doubles. Say, *When both addends in a number sentence are the same number, it is an addition **doubles fact**. Addends are the numbers being added together.*

Write the number sentences $2 + 2 = 4$ and $4 - 2 = 2$ on a chart. Say, *Two plus two equals four. This is a doubles fact because both addends are two. Four minus two equals two. This is a related fact because it uses the same numbers.* Write the phrase related facts.

Model doubles addition number sentences, such as: $3 + 3 = 6$; and $4 + 4 = 8$. Have students identify the related subtraction number sentence using the sentence frames:
____ plus ____ equals ____. So ____ minus ____ equals ____.
Have students utilize the sentence frames to report back to you or a peer Exercises 5–14 in On My Own.

English Language Development Leveled Activities

Emerging Level	Expanding Level	Bridging Level
Activate Prior Knowledge Review the word *doubles* and the Spanish cognate, *dobles*. Create two groups of 3 students. Count each group and count the total of the two groups combined. Write $3 + 3 = 6$. Say, *The addends are the same. This is an addition doubles fact.* Have one group of 3 sit back down. Count the remaining group. Write $6 - 3 = 3$. Say, *The difference is the same as the number being taken away. This is the related subtraction fact.* Repeat with different doubles facts.	**Act It Out** Position 8 students in two groups of 4. Ask each group separately, *How many are in your group?* **4** Ask, *What is the number sentence to add these two groups?* **$4 + 4 = 8$** Write $4 + 4 = 8$. Say, *Four plus four equals eight is a doubles addition fact.* Then write $8 - 4 = $ ____, and have four students sit down. Ask, *How many are left?* **4** Repeat with two groups of three. Discuss how to use doubles to subtract.	**Number Recognition** Distribute 0–10 number cards to each student. Have students pair up, shuffle their cards, and place their cards facedown in a pile. Each student turns over the top card, simultaneously. If the numbers match, students say, **Doubles!**, set the match aside, and write the addition doubles fact and the related subtraction fact. If the numbers do not match, students place the card at the bottom of their deck. Continue with the next card in each pile. The first pair to match five sets of facts, wins.

Teacher Notes:

NAME _____ DATE _____

Lesson 3 Four-Square Vocabulary
Use Doubles to Subtract

Trace the word. Write the definition for *doubles*.
Write what the word means, draw a picture, and
write your own sentence using the word.

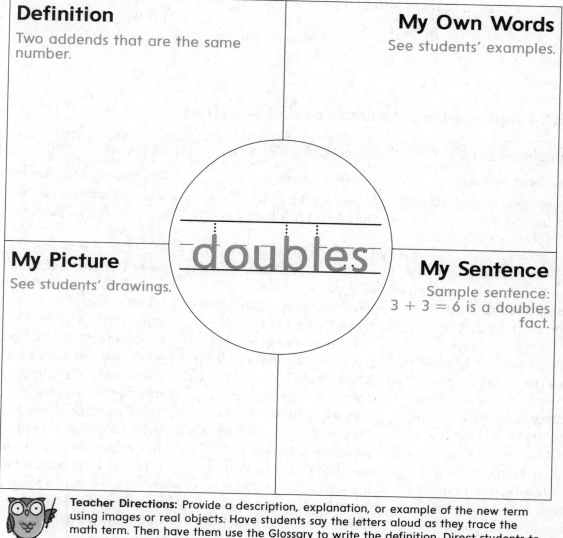

Definition
Two addends that are the same number.

My Own Words
See students' examples.

My Picture
See students' drawings.

doubles

My Sentence
Sample sentence:
3 + 3 = 6 is a doubles fact.

Teacher Directions: Provide a description, explanation, or example of the new term using images or real objects. Have students say the letters aloud as they trace the math term. Then have them use the Glossary to write the definition. Direct students to write a definition in their own words and draw a picture representing their math term. Have students write a sentence using the term and then encourage students to read their sentence to a peer.

Grade 1 • **Chapter 4** *Subtraction Strategies to 20*   **43**

Lesson 4 Problem Solving
Strategy: Write a Number Sentence
English Learner Instructional Strategy

Sensory Support: Illustrations/Diagrams

For the first Problem Solving example, have students write and display each portion of their number sentence on their write-on/wipe-off boards. Model each step as you "think aloud" writing the subtraction number sentence, $6 - 2 = 4$. Read aloud the problem. Have students echo read after each sentence. Draw a scene of 6 seagulls flying above water. Ask, *How many seagulls do you see?* **6** Write six on your board. Erase 2 seagulls and draw 2 seagulls sitting on the water. Say, *Two of the seagulls land in the ocean. Write the symbol that means to subtract.* **(−)** *Write a number to describe how many seagulls stopped flying.* **2** *Write an equals sign.* Ask, *How many seagulls are still flying?* **4** *This is what we needed to find. Write four.*

English Language Development Leveled Activities

Emerging Level	Expanding Level	Bridging Level
Listen and Write	**Number Sense**	**Developing Oral Language**
Demonstrate a subtraction story using large manipulatives. Say, *I had five cookies. I ate two. How many do I have now?* Then write $5 - 2 = $ ____. Using a number line, start at 5. Count back by 2. Say, *The difference is 3.* Have students say the answer as you write it. Distribute write-on/wipe-off boards. Model and say, *I had six marbles. I lost three. How many do I have left?* Have students use number lines, write and then display the number sentence. Repeat using realia to model other subtraction number sentences.	Model a subtraction number story using attribute buttons. Say, *I had eight buttons. Three buttons are gone. How many are left?* Have students help you write, read, and solve the subtraction number sentence for the story $8 - 3 = 5$. Divide the group into pairs and distribute at least 20 attribute buttons to each pair. Have one partner in create a subtraction number story with the manipulatives while the other partner solves the story by writing the subtraction number sentence. Have each pair present their story problem and the subtraction number sentence to another group. Then have each pair switch roles and repeat activity.	Model a subtraction number story such as, *The squirrel had twelve acorns. She ate four. How many are left?* Have students help you write, read, and solve the number sentence for the story. $(12 - 4 = 8)$. Then divide the group into pairs. Have one partner in each pair create a subtraction number story while the other partner solves the story by writing the number sentence. Have each pair retell its story and show its number sentence for the whole group. Then have each pair switch roles.

Teacher Notes:

NAME _____ DATE _____

Lesson 4 Problem Solving
STRATEGY: Write a Number Sentence

<u>Underline</u> what you know. (Circle) what you need to find. Write a subtraction number sentence to solve.

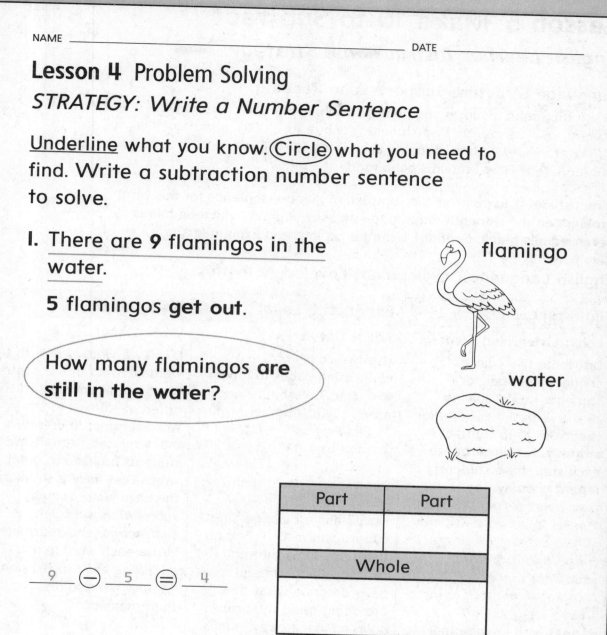

flamingo

water

I. There are **9** flamingos in the water.

5 flamingos **get out**.

How many flamingos **are still in the water?**

Part	Part
Whole	

9 ⊖ 5 ⊜ 4

__4__ flamingos **are still in the water**.

Teacher Directions: Provide a description, explanation, or example of the boldface terms and nouns using images or real objects. Read each sentence and have students echo read. Encourage students to use the part-part-whole mat and then write their answer in the restated question. Have students read the answer sentence aloud.

Lesson 5 Make 10 to Subtract
English Learner Instructional Strategy

Language Structure Support: Echo Reading

Write then read aloud as you point to each word, *There are thirteen coconuts on the island.* Have students echo read. *Seven of the coconuts roll into the ocean.* Have students echo read. *How many coconuts are left on the island?* Have students echo read.

Direct students to point at the subtraction number sentence for this word problem on the student edition page and say chorally, **Thirteen minus seven equals blank.** Continue with the Explore and Explain lesson.

English Language Development Leveled Activities

Emerging Level	Expanding Level	Bridging Level
Look, Listen, and Model	**Act It Out**	**Word Web**
Distribute two colors of connecting cubes to students. Model solving 15 − 8 with the cubes and the make 10 to subtract strategy. As you verbalize each step, have students repeat chorally. Say, *I will take apart 8 to make a 10. I need to take 5 away from 15 to make 10. Break apart 8 as 5 and 3. Subtract 5 from 15 to make 10. Then I subtract three from ten and I have 7 left. 15 − 8 − 7.* Repeat with lesson other exercises.	Use two colors of connecting cubes to model 16 − 9 = 7. Verbalize each move. *I will subtract 9 from 16. First, group 9 into 6 and 3. Then, subtract 6 from 16 to equal 10. Then, subtract 3. I have 7 left. Sixteen minus nine equals seven.* Have pairs use connecting cubes to solve 13 − 7 using the make 10 to subtract strategy. Circulate and have pairs describe what they are doing using the terms: *subtract, take away, minus,* and *equals.*	Display a word web with the word *subtract* in the center. Explain that you can add different suffixes to the word *subtract* to create many related words. Have students brainstorm other words that have a suffix on the base word, such as *subtraction, subtracts, subtracting,* and *subtracted.* Write each word and underline each suffix: *-ion, -s, -ing, -ed* and discuss their meanings.

Teacher Notes:

NAME _____ DATE _____

Lesson 5 Difference Identification
Make 10 to Subtract

Match addition examples to show how to make 10
to subtract.

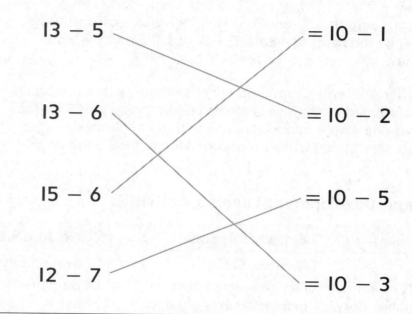

$$13 - 5 \qquad = 10 - 1$$

$$13 - 6 \qquad = 10 - 2$$

$$15 - 6 \qquad = 10 - 5$$

$$12 - 7 \qquad = 10 - 3$$

Draw a picture to show one way to make 10
to subtract.

See students' examples.

Teacher Directions: Model a subtraction sentence, such as 11 – 3, and use
manipulatives to show how to make a 10 to find the difference. Have students match
an example on the left with the make-a-10 example on the right. Direct students to
draw a picture representing a way to make 10 when subtracting. Then encourage
students to describe their picture to a peer.

Grade 1 • **Chapter 4** *Subtraction Strategies to 20* **45**

Lesson 6 Use Related Facts to Add and Subtract

English Learner Instructional Strategy

Collaborative Support: Act It Out

Write the term *related facts* on the board. Have students call out two numbers that are less than eleven, for example 8 and 7. Write the numbers on large cards. Make cards for the (+) and (=) symbols. Create an addition number sentence with the numbers and write the sum on another large card. For example 8 + 7 = 15. Say, *Using these numbers: eight, seven, and fifteen, I will create a subtraction number sentence.* Make a card with the (−) symbol. Rearrange the cards to create 15 − 7 = 8. Point to the word related facts and say, *We used **related facts** to add and subtract.*

As an extension after the lesson, provide pairs of students with 3 numbers that are in a set of related facts. Have students create three cards with the numbers, as well as one with a minus sign, one with an addition sign and one with an equals sign. Students then model making related addition and subtraction facts.

English Language Development Leveled Activities

Emerging Level	Expanding Level	Bridging Level
Act It Out Select 5 students to act out the following number story. Say, *Five children were playing.* Have students hold hands and walk in a circle. Say, *Two left.* Have two students leave the circle. *How many are left?* **3** Write 5 − 2 = 3. Continue telling the story, *Then the two came back. Now how many are there?* **5** Write 3 + 2 = 5 below 5 − 2 = 3. Draw lines connecting the like numbers and say, *These number sentences are **related facts**.* Emphasize related facts.	**Number Sense** Prepare related facts on individual strips of paper, such as 3 + 2 = 5, 5 − 2 = 3, and 5 − 3 = 2. Give one fact to each student. Have students find the other students that have related facts. Have each related fact group read their facts as you write them. Ask, *Which number is the addend in the addition sentence and the difference in the subtraction sentence?* Have students identify how each addend appears as a difference in one of the subtraction sentences.	**Developing Oral Language** Give pairs 50 counters. Have each partner take 11 counters. Have one partner model 3 + 8 = 11 and the other partner model the related subtraction fact (11 − 8 = 3 or 11 − 3 = 8). Ask, *Which addend is the difference?* Have students answer, **The addend *three/ eight* is the difference.** Have students take turns creating their own addition fact and related subtraction fact. Have each pair describe what they are doing using the words *addend, difference, add,* and *subtract*.

Teacher Notes:

NAME _____ DATE _____

Lesson 6 Note Taking
Use Related Facts to Add and Subtract

Read the question. Write words you need help with. Use your lesson to write your Cornell notes.

Building on the Essential Question	**Notes:**
How can I use related facts to add and subtract?	**Word Bank** addition check facts related facts subtraction ___Related___ ___facts___ use the same numbers. You can write related ___addition___ and ___subtraction___ facts. $2 + 5 = 7$ $7 - 5 = 2$ $5 + 2 = 7$ $7 - 2 = 5$ These ___facts___ can help you add and subtract. You can use $2 + 5 = 7$ to find $7 - 2 = 5$. You can use an addition fact to ___check___ your subtraction.
Words I need help with: See students' words.	

Teacher Directions: Read the Building on the Essential Question and have students list words/phrases they need assistance with. Provide descriptions, explanations, or examples of the terms using images or real objects. Read each sentence frame and have students write the appropriate terms. Have students read their notes aloud.

Lesson 7 Fact Families

English Learner Instructional Strategy

Collaborative Support: Act It Out

Write the term *fact family* on the board. Say, *A fact family is a group of four related addition and subtraction facts that use the same three numbers.*

Roll two 0–5 number cubes. Write the numbers on large cards. Create the number sentence for the sum of those *numbers and write the sum on another large card. For example 2 + 3 = 5. Say, Using these three numbers: two, three, and five, I will create three more number sentences.* Rearrange the cards to create 3 + 2 = 5; 5 − 2 = 3; 5 − 3 = 2.

As an extension after the lesson, provide pairs of students with 3 numbers that are used in the related facts of a fact family. Have students create three cards with the numbers, as well as one with a minus sign, one with an addition sign and one with an equals sign. Students then model making the related facts to complete the fact family.

English Language Development Leveled Activities

Emerging Level	Expanding Level	Bridging Level
Number Sense	**Word Knowledge**	**Developing Oral Language**
Show a picture of a family with parents and children. Say, *This is a family.* Write out a fact family. Say, *This is a fact family.* Emphasize *fact family.* Prepare different sets of related facts on individual strips of paper, such as 3 + 2 = 5, 2 + 3 = 5, 5 − 2 = 3, and 5 − 3 = 2. Distribute one strip of paper to each student. Say, *Find your fact family.* Then have students find the other three students that have related facts.	Review the words: *family, number fact, and fact family.* Prepare a number fact, a fact family, or a picture of a human family on numerous index cards. Shuffle and distribute one card to each student. Have students find other students with the same type of card. Have the Number Fact Group read its number facts. Have the Family Picture Group show its pictures and count the number of people in each family. Have the Fact Family Group read the fact family on each card.	Divide students into groups of four. Have one student write 13 + 8 = 21 on a piece of paper, read it aloud, and then pass it to the next student. The next student writes another number sentence in the fact family, for example 8 + 13 = 21, reads it aloud, and passes it onto the next student. Continue until all facts in the fact family have been written. Repeat the activity multiple times, allowing each student to choose the first fact to write.

Teacher Notes:

NAME _____ DATE _____

Lesson 7 Concept Web
Fact Families

Trace the words in the center oval. Draw lines to match the true examples to the term *fact family*.

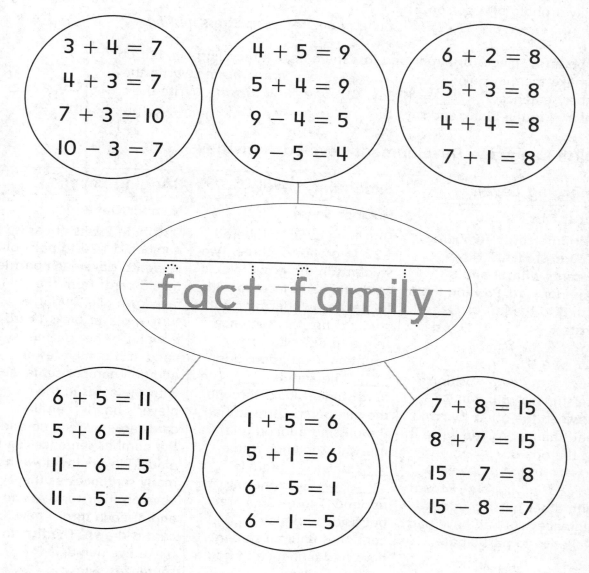

$3 + 4 = 7$
$4 + 3 = 7$
$7 + 3 = 10$
$10 - 3 = 7$

$4 + 5 = 9$
$5 + 4 = 9$
$9 - 4 = 5$
$9 - 5 = 4$

$6 + 2 = 8$
$5 + 3 = 8$
$4 + 4 = 8$
$7 + 1 = 8$

fact family

$6 + 5 = 11$
$5 + 6 = 11$
$11 - 6 = 5$
$11 - 5 = 6$

$1 + 5 = 6$
$5 + 1 = 6$
$6 - 5 = 1$
$6 - 1 = 5$

$7 + 8 = 15$
$8 + 7 = 15$
$15 - 7 = 8$
$15 - 8 = 7$

Teacher Directions: Provide a description, explanation, or example of the new term using images or real objects. Have students say the letters aloud as they trace the math term. Direct students to draw a line from each example of a fact family to the term in the center. Then encourage students to describe their work to a peer.

Grade 1 • **Chapter 4** *Subtraction Strategies to 20* **47**

Lesson 8 Missing Addends
English Learner Instructional Strategy

Language Structure Support: Multiple Word Meanings

Write the word missing. Say, *The word **missing** has multiple meanings. Let's learn about some of them.*

Pat your pockets and look around your desk. Say, *I lost my red pen. My red pen is **missing**.* Stress *missing.* Hold up a red pen and say, *I found it.* Display a photo of a person. Say, *I don't get to talk to my friend very often. I find myself **missing** the sound of his/her voice.* Stress *missing.*

Ball up a few pieces of paper and toss them near a recycling bin, but do not make them in. Say, *I keep **missing** the bin.* Stress *missing.* Write the number sentence 7 +____ = 11. Say, *The addend is **missing**.* Write the number 4 in the blank and say, *I found the **missing** addend.*

English Language Development Leveled Activities

Emerging Level	Expanding Level	Bridging Level
Act It Out	**Number Sense**	**Number Game**
Distribute cards numbered 0–20 to students. Have students with 13 and 8 come forward. Position them with 8 to the left of 13. Write 8 + ____ = 13. Ask, *What is the **missing addend?*** **5** *Who has the missing addend?* Position the student with 5 in between the other two and read the fact as you write it on the board: *Eight plus five equals thirteen. Five was the missing addend.* Repeat with other missing addend sentences until all have had a chance to participate.	Distribute cards numbered 1–20 to groups of three. Two students in the group pick a card and lay it face up with the lesser number first. Both students use this sentence frame to describe the equation: **One addend is ____. The sum is ____.** The third student looks through the numbered cards to find the missing addend and uses this sentence frame: **____ plus ____ equals ____. ____ was the missing addend.** Those cards are put aside and play continues until all students have had a chance to find a missing addend.	Distribute 2 sets of cards numbered 1–20 to pairs of students. Have each partner take 5 cards from the facedown pile. During a turn, a player picks 1 card from his or her partner's hand. If it completes an addition number sentence with the cards in the player's hand, then the cards are laid face up and the number sentence read aloud. Players lay down as many sentences as they have on their turn. At the end of each turn, 1 new card is drawn. The first to make five number sentences, wins.

Teacher Notes:

NAME _____ DATE _____

Lesson 8 Vocabulary Sentence Frames
Missing Addends

The math words in the word bank are for the sentences below. Write the words that fit in each sentence on the blank lines.

Word Bank		
count back	fact family	missing addend

1. Addition and subtraction sentences that use the same numbers like

   $7 + 8 = 15$ $15 - 8 = 7$

   $8 + 7 = 15$ $15 - 7 = 8$

   are a __fact__ __family__.

2. For $5 \oplus \underline{} = 9$, the __missing__ __addend__ is 4.

3. On a number line, start at the greater number and __count__ __back__ if you want to subtract.

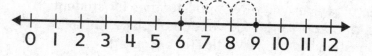

0 1 2 3 4 5 6 7 8 9 10 11 12

Teacher Directions: Provide a description, explanation, or example of the each term using images or real objects. Read each sentence frame and have students echo read. Direct students to write the correct terms in each blank. Then encourage students to read each sentence to a peer.

Chapter 5 Place Value

What's the Math in This Chapter?

Mathematical Practice 2: Reason abstractly and quantitatively

Show two piles of 27 blocks. Say, *Are these piles equal?* Count them by tens and ones. Then write "27" and "twenty-seven" on the left side of the board. Repeat on the right. Say, *Are these numbers the same?* Solicit **Yes** from students.

Say, *That means 27 is equal to 27.* Write both "=" and "equal to" between the numbers. Say, *I used numbers and words to solve the problem. I compared the numbers. They are the same, or equal.*

Ask, *Have you ever used words to solve a math problem?* Solicit **Yes** from students, then have students share or demonstrate their experiences.

Ask, *When have you used words and numbers to compare?* Have students talk with a peer then discuss as a group. The goal is to recognize that using both words and numbers can help them with comparing place value.

Display a chart with Mathematical Practice 2. Restate Mathematical Practice 2 and have students assist in rewriting it as an "I can" statement, for example: **I can use numbers and words to help me make sense of problems.** Have students draw or write examples of for each. Post the new "I can" statement and examples in the classroom.

Inquiry of the Essential Question:

How can I use place value?

Inquiry Activity Target: **Students come to a conclusion that using words and numbers helps them understand place value so they can compare numbers.**

As an introduction to the chapter, present the Essential Question to students. The inquiry graphic organizer will offer opportunities for students to observe, make inferences, and apply prior knowledge of place value representing the Essential Question. As they investigate, encourage students to draw, write, and collaborate with peers to demonstrate their observations and thinking. Then have students present additional questions they may have to a peer to extend discussions.

Regroup students and restate Mathematical Practice 2 and the Essential Question. Pose questions to reflect on what has been learned to guide students in making connections between the Mathematical Practice and the Essential Question.

NAME _____ DATE _____

Chapter 5 Place Value
Inquiry of the Essential Question:

How can I use place value?

13 equals 13

I see ...

I think ...

I know ...

34 is greater than 21

I see ...

I think ...

I know ...

22 is less than 41

I see ...

I think ...

I know ...

Questions I have...

_____

_____

_____

Teacher Directions: Read the Essential Question for students. Have students echo read. Direct students to describe their observations, inferences, and prior knowledge of each math example. Encourage students to write or draw additional questions they may have. Then have students share their ideas/questions with a peer.

Grade 1 • **Chapter 5** *Place Value* **49**

Lesson 1 Numbers 11 to 19

English Learner Instructional Strategy

Sensory Support: Physical Activities

Call out numbers in order from 1 to 10. Have students use their fingers to model each number as you say it aloud. Call out the number 11. Ask, *Can you display eleven using your fingers?* **no** Discuss why one person cannot display the number 11 with his or her 10 fingers. Call out the number 11. Have pairs use their fingers to model each number as you say it aloud. Ask, *Why can a pair display the number eleven but one person alone cannot?* Discuss how you need a group of 10 and 1 more to display 11. Continue calling out numbers in order from 12 to 19. Have pairs use their fingers to model each number.

English Language Development Leveled Activities

Emerging Level	Expanding Level	Bridging Level
Making Connections	**Sentence Frames**	**Developing Oral Language**
Write the numbers 10 to 19. Have students count from 10 to 19 in their native language and point to each number as they count. Point and count the numbers 10 to 19 as students chorally recite the number names in English. Distribute index cards labeled 1–19 to each pair of students. Have them verbally review the number on each card. Then, have pairs mix-up the cards and work together to put them in order. Circulate and assist pairs in using English number names.	Select 9 students to form a line. Distribute the same number of craft sticks as the place the student is in line (for example, 1 for the first, 2 for the second, and so on). Each student will display the craft sticks and say the number aloud (1–9). Create bundles of 10 craft sticks secured with rubber bands. Display a bundle and say, *ten*. Distribute bundles to students in the line and have each say, **Ten craft sticks and _____ more craft sticks is _____ craft sticks.**	Have students recite numbers up to 19. Then have students model counting and representing sets of numbers with their number names. For example, model 10 and 3 more. Say, *Ten and three more is thirteen.* Distribute 19 counters to each pair. Give each pair a set of index cards labeled 1–19. As students alternate displaying a card, the other student must model the number with counters. Have pairs use the sentence frames: **Show me _____ counters. Here are _____ counters.**

Teacher Notes:

NAME _____ DATE _____

Lesson I Word Web
Numbers II to 19

Trace the math word. Draw a number story in each rectangle that shows the meaning of *more*.

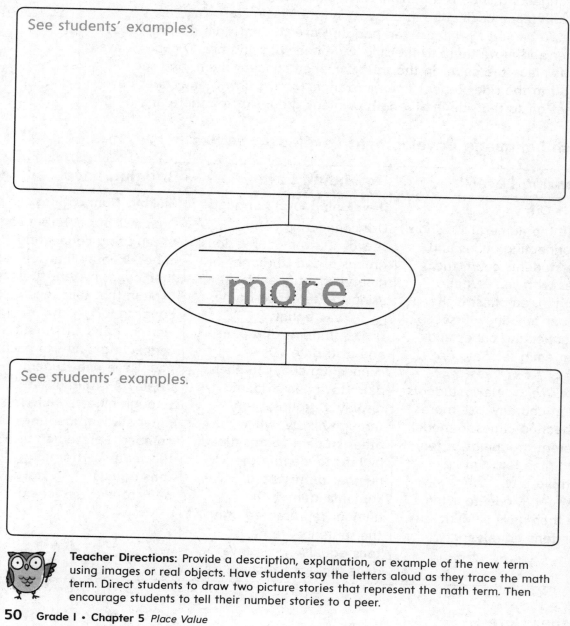

See students' examples.

more

See students' examples.

Teacher Directions: Provide a description, explanation, or example of the new term using images or real objects. Have students say the letters aloud as they trace the math term. Direct students to draw two picture stories that represent the math term. Then encourage students to tell their number stories to a peer.

Lesson 2 Tens

English Learner Instructional Strategy

Sensory Support: Hands-On Activity

Prepare "train cars" for each pair of students by removing the lid and cutting off one column from an egg carton, leaving ten spaces. The spaces will represent "seats" in the "train car." Distribute one "train car" and 90 connecting cubes to each pair of students.

Explain that the egg carton is a train car, each train car has 10 seats, and each connecting cube is a person. Have pairs count out 10 connecting cubes and place each cube in a "seat" on the "train car." *How many people are in your train car?* **ten people** Have students join the ten cubes together and move them to the side. Continue with another 10 cubes. Have students place the cubes in the train car, then connect them, and set them off to the side. Repeat one more time. Discuss how many cubes have been set off to the side in all and how many groups of ten there are in all.

English Language Development Leveled Activities

Emerging Level	Expanding Level	Bridging Level
Act It Out	**Developing Oral Language**	**Making Connections**
Write the numeral 10. Show 10 connecting cubes and have students count the cubes with you. When counting, emphasize the number ten. Show 3 sets of 10 connecting cubes and count each set. Say, *1 ten, 2 tens, 3 tens. Three tens equals thirty.* Have students repeat chorally. Use more connecting cubes to model different groupings of ten, up to nine tens. For example, say, *I will show four tens.* Model four tens, and then have students say, **Four tens equals forty.**	Use connecting cubes, in sets of 10, to model various groups of ten (through ninety). Identify each set using the sentence frame: _____ **tens equals** _____. Divide students into pairs and distribute 90 connecting cubes to each pair. Have one student display a group of ten through ninety, while the other student counts aloud by tens to identify the number represented. Have students identify the number represented using the sentence frame: _____ **tens equals** _____. Have pairs repeat, switching roles.	Demonstrate different sets of ten using connecting cubes. Identify four tens and then count by tens to forty. Explain that four tens is equal to forty. Have students work in pairs. Distribute connecting cubes to each pair. Have one student display a group of ten through ninety, while the other student identifies the number represented using the sentence frame: _____ **tens equals** _____. Have pairs repeat, switching roles.

Teacher Notes:

NAME _____ DATE _____

Lesson 2 Number Identification
Tens

Trace each word. Then draw lines to match. The first one is done for you.

Number	Tens	Word
20	4 tens	forty
40	5 tens	seventy
70	2 tens	fifty
50	8 tens	twenty
80	7 tens	eighty
60	1 ten	ten
30	6 tens	ninety
10	9 tens	thirty
90	3 tens	sixty

Teacher Directions: Use manipulatives to model each number. Model and then practice counting by tens as a group and then individually. First, have students trace each word, saying each letter as they write it. Have students say a number in the Number column, identify the matching number in the Tens column, and then draw a line to match the quantities. Students then match the Tens to the number word. Encourage partners to report about a number using a sentence frame such as: **Ninety is nine tens.**

Grade I • **Chapter 5** *Place Value* **51**

Lesson 3 Count by Tens Using Dimes

English Learner Instructional Strategy

Collaborative Support: Partners Work

Have students work in pairs during the Explore and Explain Activity. Before you begin, model how to *count by tens* using dimes. Count out 3 dimes and then *count by tens* to 30. Write 30 on the board. Distribute 4 dimes to pairs of students. Ask, *How many dimes?* Have students chorally answer, **4 dimes**. Then ask, *How much is a dime worth?* **ten cents** Ask, *What is the total worth of all the dimes?* **40 cents**. Ask, *Four tens is equal to how many ones?* **forty ones** *Four dimes is equal to how much in all?* **forty cents** Write 40 on the board.

English Language Development Leveled Activities

Emerging Level	Expanding Level	Bridging Level
Exploring Language Structure Write both the singular and plural form of the following words on the board: *coin/coins, penny/pennies, cent/cents* and *dime/dimes*. Demonstrate the value of pennies and dimes by showing that ten pennies is equal to one dime. Use money to model and say, *This is one* **coin**. *Here are two* **coins**. *One penny is one* **cent**. *Two* **pennies** *is two* **cents**. *One* **dime** *is ten cents. Two* **dimes** *is twenty cents*. Point to the singular or plural form of the listed words as you say each. Have students recite each.	**Build Background Knowledge** Distribute 10 pennies and 1 dime to students. Say, *One penny* **equals** *one cent. Two pennies* **equals** *two cents. When amounts are equal, they are the same. Two pennies are the* **same amount** *as two cents.* Have students repeat each sentence and hold up one or two pennies. Say, *Ten pennies* **equals** *one dime. A dime* **is equal to** *ten pennies. Ten pennies* **is the same amount** *as one dime.* Have students count out 10 pennies and repeat each sentence as they point to the coins.	**Act It Out** Display 10 pennies and 1 dime. Say, *Ten pennies equals one dime. Ten pennies is the same amount as one dime.* Use dimes to model numbers 10, 20, 40, 70, identifying each set using the sentence frames: ____ *dimes is equal to* ____ *tens* and so on. Distribute 9 manipulative dimes to each student. Say, *Show three tens*. Then have them model using dimes and the sentence frame: ____ **dimes is equal to** ____ **tens.** Repeat with different numbers.

Multicultural Teacher Tip

Because many word problems involve prices and/or determining changes in monetary value, ELs will benefit from an increased understanding of American coins and bills. A chart or other kind of graphic organizer visually comparing coin and bill values and modeling how to write dollars and cents in decimal form would help these students. You may also want to have ELs describe the monetary systems of their native countries. Identifying similarities or differences with the American system can help familiarize students with dollars and cents.

NAME _____ DATE _____

Lesson 3 Four-Square Vocabulary
Count by Tens Using Dimes

Trace the word. Write the definition for *dime*.
Write what the word means, draw a picture, and
write your own sentence using the word.

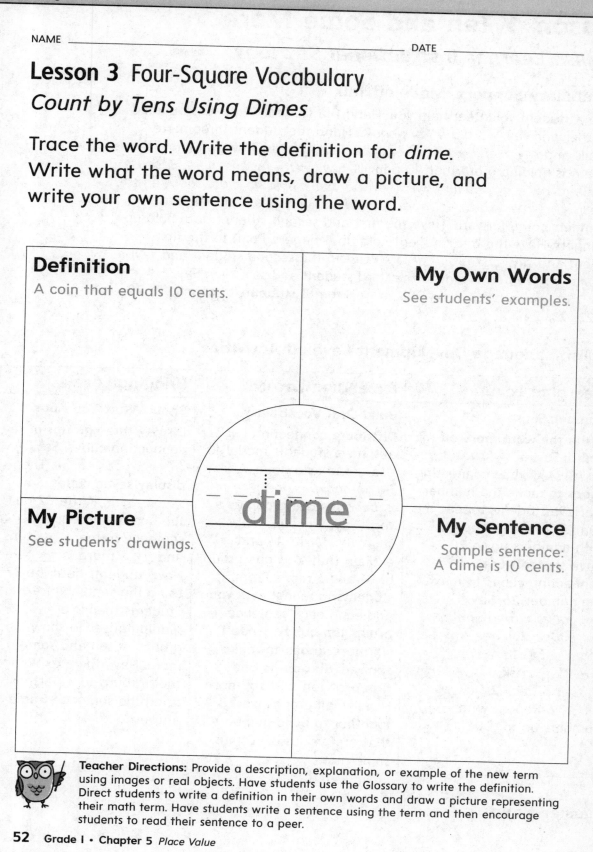

Definition
A coin that equals 10 cents.

My Own Words
See students' examples.

dime

My Picture
See students' drawings.

My Sentence
Sample sentence:
A dime is 10 cents.

Teacher Directions: Provide a description, explanation, or example of the new term using images or real objects. Have students use the Glossary to write the definition. Direct students to write a definition in their own words and draw a picture representing their math term. Have students write a sentence using the term and then encourage students to read their sentence to a peer.

Lesson 4 Ten and Some More
English Learner Instructional Strategy

Vocabulary Support: Modeled Talk

Have a student volunteer help you. Hand the student 10 pencils bundled together and say, *Here are ten pencils.* Hand the student three more pencils and say, *Here are three extra pencils.* Discuss how you know the student is holding more than 10 pencils. Say, *extra means more. Because I gave him/her extra pencils, we know he/she is holding more than ten pencils. One ten and three more is thirteen.* Have two more student volunteers come forward. Have the first and second student both hold up 10 fingers. Have the third student hold up 4 fingers. Point to the first student and say, *one group of ten.* Point to the second student and say, *a second group of ten.* Point to the third student and say, *four.* Say, *Here we have two groups of ten and four more.* Prompt students to count, **10, 20, 21, 22, 23, 24.**

English Language Development Leveled Activities

Emerging Level	Expanding Level	Bridging Level
Number Sense Write the word more on the board. Say, *I will show 1 ten and 3 more.* Use connecting cubes to make the number 13. Have students repeat the number chorally. Repeat with numbers up to 20. Have students use connecting cubes to make the number 26. Say, *Group ten cubes. Group another ten cubes. You need six more cubes to make twenty-six.* Ask, *How many groups of ten?* **2** *How many more?* **6** Repeat with numbers up to 100.	**Build Oral Vocabulary** Distribute connecting cubes, and have students model 10. Use group as a noun. Ask, *Do you have a group of ten cubes?* **yes** Use group as a verb. Say, *I group cubes to show ten.* Write 16 and explain that 16 is one group of 10 and 6 more. Prompt students to model with you and echo each sentence: **I group ten cubes. I need six more cubes to make sixteen. Sixteen is one group of ten and six more.** Have small groups work together to build numbers that are groups of 10 plus some more.	**Show What You Know** Display three groups of ten connecting cubes. Say, *three groups of ten.* Then display seven cubes. Say, *seven more.* Write 37 on the board. Introduce the sentence frame: **____ tens and ____ more is ____.** Have students describe 37 using the sentence frame. Students then use manipulatives to show groups of ten and some more. Have them explain their groups to a partner using the sentence frame above.

Teacher Notes:

NAME _____ DATE _____

Lesson 4 Note Taking
Tens and Some More

Read the question. Write words you need help with. Use your lesson to write your Cornell notes. Write or draw math examples to explain your thinking.

Building on the Essential Question	Notes:
How can I count ten and some more?	 First, I circle groups of __ten__. Then, I __count__ the groups of ten. Next, I count the __ones__. Last, I write the __numbers__. __3__ tens and __2__ ones is __32__.

Words I need help with:
See students' words.

My Math Examples:
See students' examples.

Teacher Directions: Read the Building on the Essential Question and have students list words/phrases they need assistance with. Provide descriptions, explanations, or examples of the terms using images or real objects. Read each sentence frame and have students write the appropriate terms. Have students read their notes aloud. Direct students to draw a picture representing the question. Then encourage students to describe their picture to a peer.

Grade 1 • **Chapter 5** *Place Value* **53**

Lesson 5 Tens and Ones
English Learner Instructional Strategy

Sensory Support: Visual Models

Prepare 2 egg cartons for this activity by removing the lids and cutting off one column from each egg carton, leaving ten spaces. Display 23 counters in a pile. Have students count with you as you count all 23 counters. Say, *How many ones?* **23** *We can regroup the ones to create groups of ten.* Have student volunteers help you put one counter in each egg carton space. Once an egg carton is full, say, *We put ten ones in this carton. Ten ones is the same as one ten. This carton holds one ten.* Move the carton to the side and fill another carton with 10 counters. Say, *How many groups of ten?* **2** *The counters left over are ones. How many ones?* **3** Continue with the Explore and Explain Activity.

English Language Development Leveled Activities

Emerging Level	Expanding Level	Bridging Level
Activating Prior Knowledge	**Show What You Know**	**Making Connections**
Show ten individual connecting cubes, unconnected. Say, *Here are ten cubes. Each cube is one. This is ten ones.* Write: 10 ones. Tell students, *You can regroup ten ones as one ten.* Connect the cubes to create a cube train of ten. Write: = 1 ten after 10 ones on the board then say, *Ten ones equals one ten.* Distribute 13 cubes to each student. Prompt them to show one ten and three more. Say, *You regrouped to show 1 ten and three ones. Say regroup.* **regroup**	Write: 20 ones. Then show 20 individual connecting cubes. Count each cube, up to twenty. Have students echo count. Next, write: 2 tens. Show two groups of ten connecting cubes. Count the groups by tens; 10, 20. **10, 20** Have students work in pairs to model groups of ten up to 90. One student names the group of ten (up to 90) and the other student models the number using connecting cubes. Students check their partner's model for accuracy by counting aloud.	Group students in multilingual teams. Have each group represent tens and ones in many different ways as possible. Suggest that teams use models, actions, native language, or pictures. For example, the number 34 can be represented using words (Three tens and four ones), by drawing a picture (3 groups of 10 stars and 4 individual stars), using fingers (3 students using all 10 fingers and 1 student using 4 fingers), or modeled with cubes (3 ten-trains and 4 individual cubes). Ask each group to discuss their representations.

Teacher Notes:

NAME _____ DATE _____

Lesson 5 Word Web

Tens and Ones

Use the word web to show examples of regrouping.

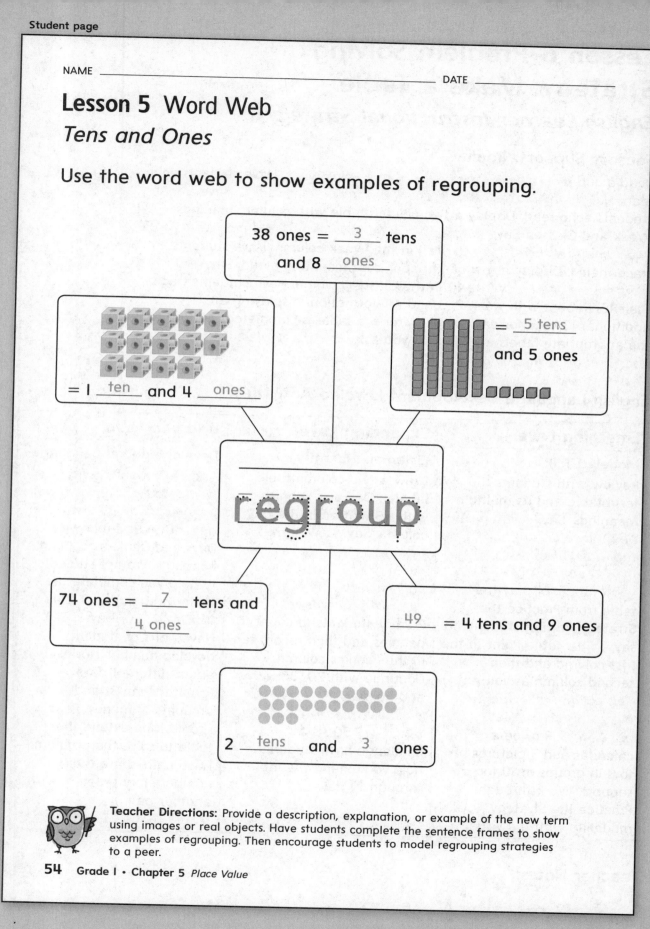

38 ones = __3__ tens
and 8 __ones__

= 1 __ten__ and 4 __ones__

= __5 tens__
and 5 ones

regroup

74 ones = __7__ tens and
__4 ones__

__49__ = 4 tens and 9 ones

2 __tens__ and __3__ ones

Teacher Directions: Provide a description, explanation, or example of the new term using images or real objects. Have students complete the sentence frames to show examples of regrouping. Then encourage students to model regrouping strategies to a peer.

Lesson 6 Problem Solving Strategy: Make a Table

English Learner Instructional Strategy

Sensory Support: Realia

Read aloud, *Joel drinks ten glasses of milk each week.* Have students echo read. *How many glasses of milk does he drink in four weeks?* Have students echo read. Display a two-column table with column headings: Week and Glasses. Say, *We can use a **table** to record the information we know and find the answer.* Write 1 in the Week column. Have students trace their 1. Display a row of 10 paper cups, counting aloud. Say, *Ten ones equals one ten.* Write 10 in the Glasses column. Have students trace their 10. Repeat with week 2, 3 and 4 information, displaying an additional 10 cups for each week. After the table is completed, gesture to the appropriate labels and data as you ask, *How many glasses of milk does Joel drink in four weeks?* **40**

English Language Development Leveled Activities

Emerging Level	Expanding Level	Bridging Level
Modeled Talk Review with students the term table and its multiple meanings. Say, *A **table** can be a piece of furniture. A **table** can also be a chart to help keep track of information.* Display the table from Practice the Strategy. Prompt students to say, *table.* **table** Point to the first column and then second column as you say, *This column tells about **days**. This column tells about **toys**.* Provide a calendar and 5 pictures of toys in groups of 10 for support. Work through Practice the Strategy modeling with the visuals.	**Listen and Identify** Draw a two-column table labeled Day and Pennies with 1–5 written in the Day column. Say, *How many pennies would I have after five days if I saved ten pennies a day? A table can help with the answer.* Prompt students to count 10 pennies and then record ten in the Pennies column. Continue with 20, 30, 40, and 50. Ask, *How many pennies would I have after five days?* **50** As a challenge have students help you extend the table through 10 days.	**Developing Oral Language** Ask, *How much money would I have if I saved ten pennies a day for three days?* Draw a table with Days and Cents as column headings. Prompt students to help you complete the table through day 3 (Day 1, 10; Day 2, 20; Day 3, 30). Have pairs of students develop number stories about different days (greater than 3) and amounts of money. Pairs should then extend the table to solve the problem. Have pairs share their number story problem with another pair.

Teacher Notes:

NAME _____ DATE _____

Lesson 6 Problem Solving
STRATEGY: Make a Table

<u>Underline</u> what you know. (Circle) what you need to find. Make a table to solve.

1. **Mara** has 3 **groups** of 10 balls.

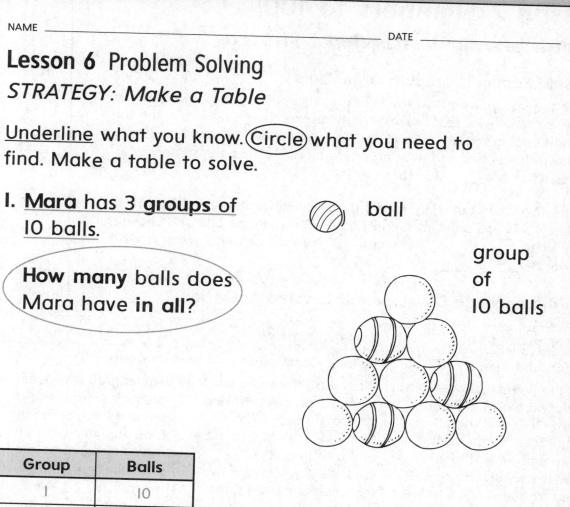

ball

group of 10 balls

(How many balls does Mara have **in all**?)

Group	Balls
1	10
2	20
3	30

Mara has __30__ balls in all.

Teacher Directions: Provide a description, explanation, or example of the boldface terms and nouns using images or real objects. Read each sentence and have students echo read. Encourage students to use the table to organize the information and then write their answer in the restated question. Have students read the answer sentence aloud.

Grade I • **Chapter 5** *Place Value* **55**

Lesson 7 Numbers to 100

English Learner Instructional Strategy

Graphic Support: Graphic Organizers

Before the lesson, frontload the following vocabulary using photos or realia: *block/blocks, leaf/leaves*. Briefly discuss the plural forms and the different ending sounds /s/ and /z/. To check understanding, say a term and have them point to the appropriate picture or use the sentence frames: **That is a ____.** or **Those are ____.**

For the Talk Math question, model the following sentence frame with the number 42 as you point to supporting examples in the See and Show:
I can write ____ as ____ tens and ____ ones. I can also write it with the words ____ – ____.

English Language Development Leveled Activities

Emerging Level	Expanding Level	Bridging Level
Listen and Identify	**Making Connections**	**Synthesis**
On the board, draw a house with two rooms. Label one room *Ones* and the other room *Tens*. Name each room and have students repeat chorally. Write 63. Point at 6 and say, *six tens.* Point to the Tens room. Say, *Six is in the tens room.* Point at 3 and say, *three ones.* Point to the Ones room. Say, *Three is in the ones room.* Repeat with other two-digit numbers. Have students come to the board and point to the room each digit is in.	Display a tens and ones place value chart. Provide students a copy of Work Mat 7. Distribute unit cubes and rods to pairs of students. Say aloud a two-digit number. Have pairs work together to write and show the number, placing the manipulatives in the correct columns of the place-value chart. Have students say, ____ **ones and** ____ **tens is** ____. Have a volunteer come to the front and write the number in the place value chart. Repeat with other two-digit numbers.	Model groups of tens and ones for several two-digit numbers. For each number, have students write the number and complete the sentence frame: **This number is ____ tens and ____ ones.** Divide students into pairs. Have partners take turns explaining how two-digit numbers can be written differently. As one student shares each explanation, the other student should write the number in words, the groups of tens and ones, or the number described as tens and ones.

Teacher Notes:

NAME _____ DATE _____

Lesson 7 Number Identification
Numbers to 100

Trace each word. Then draw lines to match. The first one is done for you.

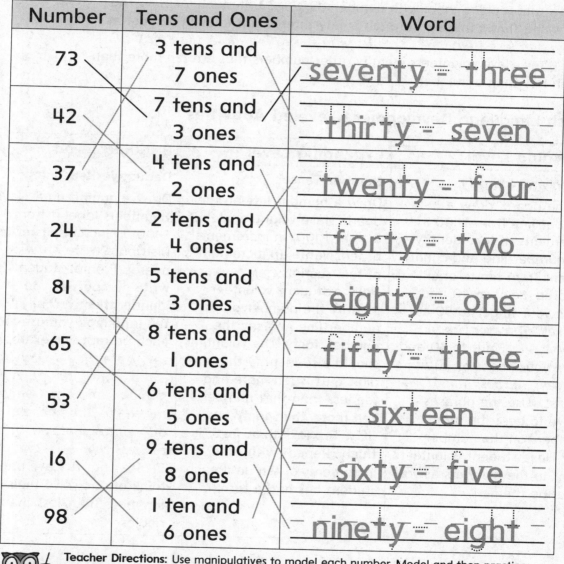

Number	Tens and Ones	Word
73	3 tens and 7 ones	seventy - three
42	7 tens and 3 ones	thirty - seven
37	4 tens and 2 ones	twenty - four
24	2 tens and 4 ones	forty - two
81	5 tens and 3 ones	eighty - one
65	8 tens and 1 ones	fifty - three
53	6 tens and 5 ones	sixteen
16	9 tens and 8 ones	sixty - five
98	1 ten and 6 ones	ninety - eight

Teacher Directions: Use manipulatives to model each number. Model and then practice counting by tens as a group and then individually. First, have students trace each word, saying each letter as they write it. Have students say a number in the Number column, identify the matching number in the Tens column, and then draw a line to match the quantities. Students then match the Tens and Ones to the number word. Encourage partners to report about a number using a sentence frame such as: **37 is three tens and seven ones.**

Lesson 8 Ten More, Ten Less

English Learner Instructional Strategy

Graphic Support: Number Lines

Create a 10–90 number line using a long piece of masking tape and student-held pieces of paper numbered 10, 20, 30, through 90. Say, *The number line shows ten through ninety.* Point to each number as you say, **Ten less than** *forty is thirty.* **Ten more than** *forty is fifty.* Repeat with other examples. Encourage students to echo your descriptions. Collect the numbers and replace them with numbers 13, 23, 33, through 93. Say, The number line shows thirteen through ninety-three. Point to each number as you say, **Ten less than** *forty-three is thirty-three.* **Ten more than** *forty-three is fifty-three.* Have volunteers identify other numbers that are ten more than and ten less than a number.

English Language Development Leveled Activities

Emerging Level	Expanding Level	Bridging Level
Listen and Identify	**Number Sense**	**Deductive Reasoning**
On the board, draw a number line from 1–100 in increments of ten. Identify the number line and count by tens from ten to one hundred. Point to 40 and move your finger to 50 as you say, *Ten more than forty is fifty.* Point to 40 and move your finger to 30 as you say, *Ten less than forty is thirty.* Use the phrases above to describe other numbers on the number line. Have student volunteers point to the numbers as you describe them.	Draw a number line on the board. Have volunteers label it from 1–100, in increments of ten. Name an increment of ten, such as *four tens.* Have a student volunteer identify the number on the number line and say the number, **forty.** The volunteer should also identify the tens digit that is ten more and ten less than that number, **Ten more than forty is fifty. Ten less than forty is thirty.** Repeat with other increments of ten until each student has had a turn.	Draw a number line. Have volunteers label it from 1–100, in increments of ten. Distribute write-on/wipe-off boards. Prompt students to write numbers as you identify them verbally. Vary the language you use for each prompt. For example, say, *I am the number that is ten more than 30. What number am I?* or *I am four tens. What number am I?* or *I am the number that is ten less than 50. What number am I?* Have pairs think of and solve their own ten more than/less than riddles.

Teacher Notes:

NAME _____ DATE _____

Lesson 8 Concept Web
Ten More, Ten Less

Trace the words. Then look at the number in each box. Write the number that is 10 more or 10 less.

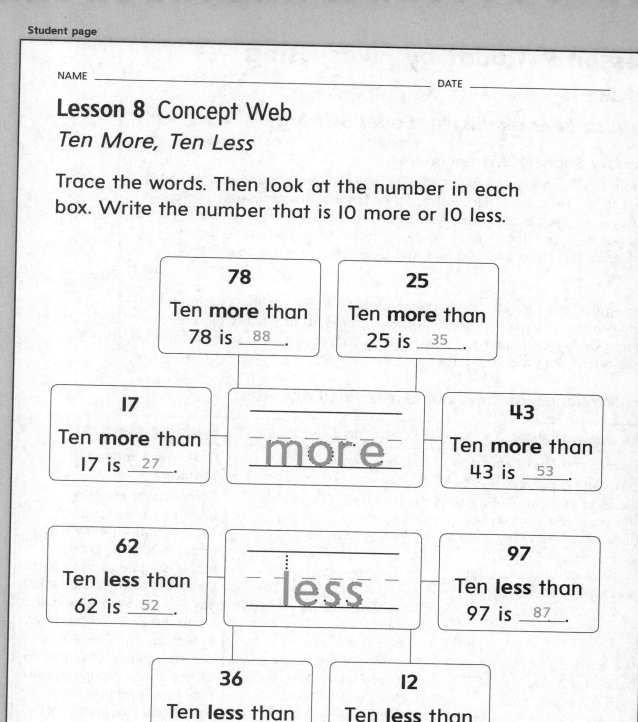

78

Ten **more** than 78 is __88__.

25

Ten **more** than 25 is __35__.

17

Ten **more** than 17 is __27__.

_____

more

43

Ten **more** than 43 is __53__.

62

Ten **less** than 62 is __52__.

_____

less

97

Ten **less** than 97 is __87__.

36

Ten **less** than 36 is __26__.

12

Ten **less** than 12 is __2__.

Teacher Directions: Provide math and non-math descriptions, explanations, or examples of the terms using images or real objects. Have students say each letter as they trace the term. Then have students complete the sentences in the web by writing a number that is ten more or ten less than the given number. Have students read their completed sentences to a peer.

Grade 1 • **Chapter 5** *Place Value* **57**

Lesson 9 Count by Fives Using Nickels

English Learner Instructional Strategy

Sensory Support: Manipulatives

Before the Explore and Explain Activity, have students count with you by ones to 50. Then use a demonstration hundred chart and have them count with you by fives to 50 again. Ask, *Did we get to 50 faster counting by ones or by fives?* **fives** Then model how to *count by fives* using nickels. Distribute nickels to each student and have them practice counting by fives.

During the activity, have students use a nickel to help them distinguish between the other coins' characteristics in the bag to identify nickels. During On My Own, pair ELLs with native English speaking students and have them count aloud together for Exercises 3–5.

English Language Development Leveled Activities

Emerging Level	Expanding Level	Bridging Level
Developing Oral Language Distribute 5 pennies and 1 nickel to students. Point out the word one cent on the back of a penny. Say, *One penny **equals** one cent.* Have students repeat. Say and have students echo, *Two pennies **equals** two cents.* Say, *Five pennies **equals** five cents.* Point out the words five cents on the back of a nickel. Say, *A nickel **is equal to** five pennies. Five pennies **equals** one nickel.* Have students count out 5 pennies and repeat each as they point to the coins.	**Number Sense** Display 4 nickels. Say, *One nickel equals five cents.* Demonstrate counting nickels by fives up to 20. Say, *five, ten, fifteen, twenty.* Ask, *How many cents is four nickels?* **20 cents** Distribute 10 nickels to each student. Say, *One nickel **equals** five cents.* We want to find out how many cents are ten nickels. Have students count the 10 nickels by fives up to 50. Ask, *How many cents is ten nickels?* **50**	**Exploring Language Structure** Demonstrate counting nickels by fives up to 50. Explain that a penny is equal to a cent, and a nickel is equal to five cents. Distribute coins to students. Say, *30* and have students count out by 5s to reach 30. (6 nickels) Then have them describe their coins in terms of how many groups of five they have and how many cents total (six groups of five, six nickels, thirty cents). Repeat until students are confident.

Teacher Notes:

NAME _____ DATE _____

Lesson 9 Four-Square Vocabulary
Count by Fives Using Nickels

Trace the word. Write the definition for *nickel*.
Write what the word means, draw a picture, and
write your own sentence using the word.

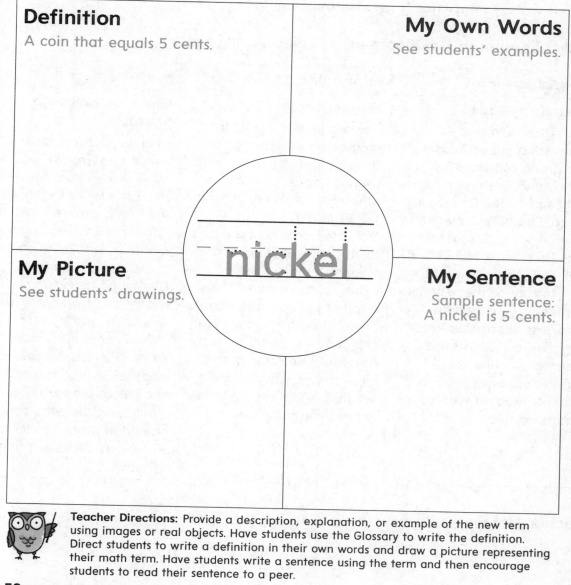

Definition
A coin that equals 5 cents.

My Own Words
See students' examples.

nickel

My Picture
See students' drawings.

My Sentence
Sample sentence:
A nickel is 5 cents.

Teacher Directions: Provide a description, explanation, or example of the new term using images or real objects. Have students use the Glossary to write the definition. Direct students to write a definition in their own words and draw a picture representing their math term. Have students write a sentence using the term and then encourage students to read their sentence to a peer.

Lesson 10 Use Models to Compare Numbers

English Learner Instructional Strategy

Vocabulary Support: Anchor Chart

Before the lesson, have ELs assist in making an Anchor Chart to frontload the lesson vocabulary. Write 13 and 19. Have students help you model each number with connecting cubes then draw the model on the chart. Say, *Thirteen is less than nineteen.* Assist students with writing the sentence. Repeat with the numbers 24 and 24. Having students write: **Twenty four is equal to twenty-four.** Repeat with the numbers 25 and 18. Having students write: **Twenty five is greater than eighteen.**

English Language Development Leveled Activities

Emerging Level	Expanding Level	Bridging Level
Word Knowledge Use yellow and red connecting cubes. Model one group of five yellow cubes, and one group of eight red cubes. Count each group. Say, *Five is less than eight. Eight is greater than five.* Then show two sets of four. Say, *Four is equal to four.* Introduce the sentence frames: ____ **is greater than/less than/equal to** ____. Have students use them to describe the number of cubes in your example. Repeat with other groups up to 10.	**Number Sense** Use connecting cubes to model and describe sets that are *greater than, less than,* or *equal to* each other. For example, 25 and 17; 24 and 32; 16 and 16. Distribute connecting cubes to students. Have pairs of students create a set of cubes that is greater than another set and describe it. Then have pairs create two sets that are equal to each other and describe them. Finally have pairs create a set that is less than another set and describe it.	**Exploring Language Structure** Model synonyms for the words: *greater (more), less (fewer),* and *equal (same as).* Explain that in math the words *greater, less,* and *equal* are used to compare numbers. Distribute counters to pairs of students. Challenge pairs to create examples of *greater than, less than,* and *equal to.* When completed, have pairs describe the sets to another pair of students or the whole group. When describing the sets, pairs should utilize math terms as well as corresponding synonyms.

Teacher Notes:

NAME _____ DATE _____

Lesson 10 Symbol Identification
Use Models to Compare Numbers

Match each symbol to its meaning.

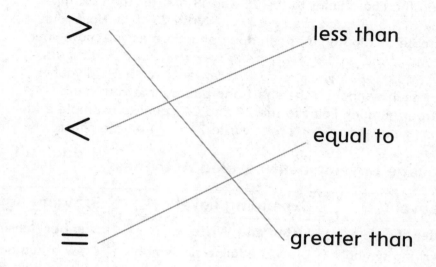

>   less than

<   equal to

=   greater than

Write the correct symbol meaning from above for each sentence on the blank lines.

1. 72 > 31   72 is __greater__ __than__ 31.

2. 50 = 50   50 is __equal__ __to__ 50.

3. 18 < 44   18 is __less__ __than__ 44.

 Teacher Directions: Review the symbols to compare numbers using images or real objects. Have students say each symbol meaning then draw a line to match the symbol to its meaning. Direct students say each number comparison and then write the corresponding meanings in the sentences. Encourage students to read the sentences to a peer.

Grade 1 · Chapter 5 *Place Value* **59**

Lesson 11 Use Symbols to Compare Numbers

English Learner Instructional Strategy

Sensory Support: Memory Devices

Add the symbols (>, <, =) to the Anchor Chart made for Lesson 10. Point to the 13 and 19 less than example and say, *Thirteen is less than nineteen.* Write 13 < 19 on the chart. Point to the 25 and 18 greater than example and say, *Twenty-five is greater than eighteen.* Write 25 > 18. Hold your hand up and create a "mouth" that can open and close as you move your fingers. Say, *Think of the greater than and less than symbol as a mouth. It is a very hungry mouth and wants to eat the larger number.* Hold your hand in front of each symbol on the chart and pantomime your hand "eating" the larger number. Point to the 24 and 24 equal to example and say, *Twenty-four is equal to twenty-four.* Write 24 = 24 on the chart.

English Language Development Leveled Activities

Emerging Level	Expanding Level	Bridging Level
Listen and Identify Draw the following symbols on the board: >, <, =. Then write the following sets of numbers, with circles between each pair: 34 ○ 68; 42 ○ 42; 72 ○ 12. Say, *Thirty-four is less than sixty-eight.* Write < between the numbers. Say, *Forty two is equal to forty two.* Write = between the numbers. Say, *Seventy two is greater than twelve.* Write > between the numbers. Repeat with other number sets. Have students show thumbs-up or thumbs down as you insert symbols between each set.	**Listen and Write** Draw the following symbols: >, <, =. Use a number line to demonstrate comparing numbers. For example: 8 is greater than 4, 5 is less than 8, and 6 is equal to 6. Use symbols to write the expressions: 8 > 4, 5 < 8, and 6 = 6. Prompt students to write different number comparisons as you say them, such as: *Twelve is greater than six (12 > 6); Eight is less than twelve (8 < 12); Four is equal to four (4 = 4).*	**Number Sense** Draw a number line from 1–20. Explain that the symbols <, >, and = stand for the terms: less than, greater than, and equal to. Use the number line to compare 10 and 14. Say, Ten *is less than* fourteen. *Fourteen is greater than* ten. Beneath the number line, write 10 < 14 and 14 > 10. Next, compare 18 and 18 and write 18 = 18. Have pairs write three of their own number comparisons using symbols. Have pairs describe their comparisons to the class.

Teacher Notes:

NAME _____ DATE _____

Lesson II Vocabulary Sentence Frames
Use Symbols to Compare Numbers

The math words and symbols in the word bank
are for the sentences below. Write the words that
fit in each sentence on the blank lines. Write the
correct symbol between the pictures.

Word Bank
equal to (=)   less than (<)   greater than (>)

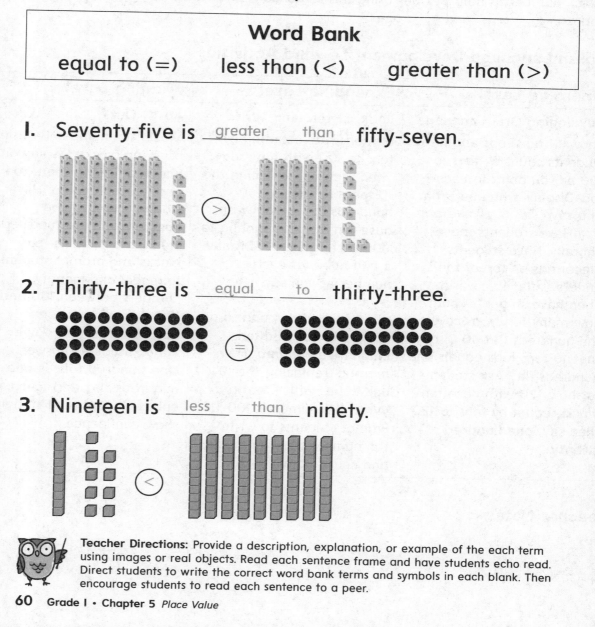

I. Seventy-five is __greater__ __than__ fifty-seven.

2. Thirty-three is __equal__ __to__ thirty-three.

3. Nineteen is __less__ __than__ ninety.

Teacher Directions: Provide a description, explanation, or example of the each term
using images or real objects. Read each sentence frame and have students echo read.
Direct students to write the correct word bank terms and symbols in each blank. Then
encourage students to read each sentence to a peer.

Lesson 12 Numbers to 120
English Learner Instructional Strategy

Vocabulary Support: Modeled Talk

Model the number 99 using the base-ten blocks in the Virtual Manipulatives online. As you talk students through the Explore and Explain Activity, model along with them to support with a visual example.

As an extension to the lesson, have student pairs take turns using the online Virtual Manipulatives, base-ten blocks to model each number in Exercises 1–6. Have students report how many hundreds, tens, and ones are in each number to their partner using this sentence frame: **There is ____ hundred, ____ tens, and ____ ones.**

English Language Development Leveled Activities

Emerging Level	Expanding Level	Bridging Level
Developing Oral Language Show 10 tens rods and have students count by tens to 100 as you point to each rod. Display a hundreds flat and say, *This is a hundreds flat.* Have students repeat chorally. Have students place rods on top of the hundreds flat to cover it. Then have students count how many tens rods covered the hundreds flat. **10** Discuss that 10 tens rods equals one hundreds flat. Ask students to show 120 with base-ten blocks, count by tens to 120 then say, **one hundred twenty.**	**Look, Listen, and Write** Demonstrate counting by tens to 100 with ten rods. Then regroup, replacing the 10 ten rods with a 1 hundreds flat. Model with base-ten blocks, counting to 110; starting at 100. Display a unit cube for each number as you say, *One hundred, one hundred one, one hundred two* and so on, up to one hundred ten. Emphasize *hundred*. Then regroup, replacing the 10 unit cubes with 1 tens rod. Write the numerals 100–110. Prompt students to write the numerals as you model and count from 110–120.	**Act It Out** Model the number 114 using base-ten blocks. Display one hundred flat, one tens rod, and four unit cubes. Have students identify the number you have modeled. **114** Distribute base-ten blocks and prompt students to model and identify numbers between 100 and 120. Say, *Show me one hundreds flat and 4 unit cubes.* Students answer, **One hundred four is one hundreds flat and four unit cubes.** Repeat until students show confidence.

Teacher Notes:

NAME _____ DATE _____

Lesson 12 Vocabulary Word Study
Numbers to 120

Circle the correct word to complete the sentence.

1. You can use hundreds, _____, and ones to show a three-digit number.

(tens)   doubles   ones

Show what you know about the word:

hundred

There are __7__ letters.

There are __2__ vowels.

There are __5__ consonants.

__2__ vowels + __5__ consonants = __7__ letters in all.

Draw a picture to show what the word means.

See students' examples.

Teacher Directions: Provide a description, explanation, or example of the new term using images or real objects. Read the sentence and have students circle the correct word. Direct students to count the letters, vowels and consonants in the math term, then complete the addition number sentence. Guide students to draw a picture representing their math term. Then encourage students to describe their picture to a peer.

Grade 1 · Chapter 5 *Place Value* **61**

Lesson 13 Count to 120

English Learner Instructional Strategy

Graphic Support: Charts

Before the lesson, display a demonstration chart that shows the numbers 1 through 120. Point to the number 43 and have students chorally say the number. **43** Ask, *What number comes next?* **44** Repeat with the numbers 50, 102, 80, and 110. Demonstrate with the numbers 50 and 80 that the next number tracks down to the next row just when we read.

Reverse roles with the students, having one of them suggest a number for you to find on the chart. Then have the class chorally ask, ***What number comes next?*** Point to the next number and say that number.

English Language Development Leveled Activities

Emerging Level	Expanding Level	Bridging Level
Developing Oral Language Display a number chart showing numbers to 120. Say, *I will point to a number. Then I will name it. Show thumbs-up if I say the correct number. Show thumbs-down if I say the wrong number.* Point to and correctly name random numbers on the chart. Every once in a while, point to a number and say the wrong name. For example, point to 45 and say, *fifty four.* Students should show a thumbs-down.	**Number Sense** Demonstrate regrouping by trading 10 ten rods for 1 hundreds flat. Distribute base-ten blocks and write-on/wipe-off boards to pairs of students. Prompt pairs to model, describe, and write numbers as you say them. For example, *one hundred eighteen* would be modeled as 1 hundreds flat, 1 tens rod, and 8 unit cubes; described as **one hundred, one ten, and eight ones,** and written as 118. Circulate among pairs as you are calling out numbers.	**Act It Out** Model 114 using base-ten blocks. Have students identify the number modeled, write the numeral **114** on write-on/wipe-off boards, and describe the number using, hundreds, tens and ones: **one hundred, one ten, and four ones.** Distribute base-ten blocks to pairs of students. Have students take turns using the blocks to model numbers up to 120 while repeating the modeled activity. Have each pair share one of the numbers they identified with another pair.

Teacher Notes:

NAME _____ DATE _____

Lesson 13 Note Taking
Count to 120

Read the question. Write words you need help with. Use your lesson to write your Cornell notes.

Building on the Essential Question	Notes:
How can I use strategies to count to 120?	**Word Bank** chart next order I should always count in <u>order</u>. I, 2, 3, 4, 5 ~~2, 3, 5, 4, I~~ I can use a number <u>chart</u> if I need help. I say a number, and then I say the number that comes <u>next</u>.

Words I need help with:

See students' words.

Teacher Directions: Read the Building on the Essential Question and have students list words/phrases they need assistance with. Provide descriptions, explanations, or examples of the terms using images or real objects. Read each sentence frame and have students write the appropriate terms. Have students read their notes aloud.

Lesson 14 Read and Write Numbers to 120

English Learner Instructional Strategy

Vocabulary Support: Modeled Talk

Before the lesson, have an older peer review how to read and write numbers to 120. Display a number chart with numbers 1–120. Use self-sticking notes to cover up the numbers 21, 55, 108, and 113. Say, *This is a number chart. A number chart can help you count.* Point to the sticky notes and say, *There are some numbers covered up on the chart. We will find and write these numbers.* Have the student count with you from 1–120. As you say each number name, point to the numeral on the number chart. When you reach the end of a row, emphasize moving to the beginning of the next row down. When you reach a number that is covered, ask, *What is the missing number?* Have the student write the numeral on the self-sticking note and say it aloud. Continue until you have reached 120.

English Language Development Leveled Activities

Emerging Level	Expanding Level	Bridging Level
Phonemic Awareness	**Number Sense**	**Act It Out**
Distribute a hundred chart and two different colored highlighters (pink and yellow) to each student. Have students work in pairs. Direct students to highlight the number 26 in pink. Then have them highlight 62 in yellow. Say the numbers verbally and/or write them out in word form. Have pairs check each other's work. Repeat with similar numbers such as: 47 and 74, 61 and 16, 19 and 90, 13 and 30 and so on. Emphasize the word endings *-teen* (thirteen) and *-ty* (thirty).	Have students work in pairs. Distribute a number chart (to 120) and correction fluid/ tape to each student. Direct students to cover 12 random numbers on their number charts and then trade charts with their partner. Students will use their counting skills to determine what numbers should be written in the covered spaces. After students have written numbers in all the covered squares, have pairs compare their charts and practice counting the numbers one through 120 aloud in unison.	Have 10 students work together as a team. Assign each student a specific color and numeral (0–9). Teams will write numbers up to 120, starting with zero. Each student will write his or her assigned numeral every time it is needed. For example, when writing the number 37, if the student assigned 3 has a red marker, and the student assigned 7 has a brown marker, then the number 37 will have a red 3 and a brown 7. To extend the activity, teams can time how long it takes for them to get to 120, then repeat the team writing activity to beat that time.

Teacher Notes:

NAME _____ DATE _____

Lesson 14 Word Identification
Read and Write Numbers to 120

Match each word to its example.

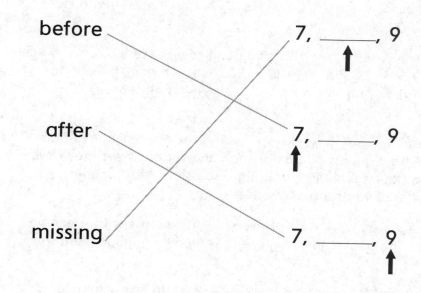

before

after

missing

7, _____, 9
↑

7, _____, 9
↑

7, _____, 9
↑

Write the correct word from above for each sentence on the blank lines.

53, _54_, 55

53 comes __before__ the missing number.

55 comes __after__ the missing number.

The __missing__ number is 54.

Teacher Directions: Review the words using images or real objects. Have students say each word and then draw a line to match the word to its example. Direct students say each word again and then complete the sentences. Encourage students to read the sentences to a peer.

Grade 1 • **Chapter 5** *Place Value* **63**

Chapter 6 Two-Digit Addition and Subtraction

What's the Math in This Chapter?

Mathematical Practice 4: Model with mathematics

Help students understand the verb form of *model*. Using raised hands of volunteers (or images of hands with all 10 fingers spread), model $10 + 10 = 20$. Say, *There are two groups of 10. There is 20 in all. We modeled adding by tens.*

Say, *I'll model more math for you.* Using pennies or base 10 blocks, model for students $14 + 9 = 23$ and $30 - 10 = 20$. As you model, be sure to describe your work saying, *I am modeling....* When finished, say, *I modeled to solve this math problem.*

Ask, *Have you ever modeled to solve a math problem?* Solicit **Yes** from students, then have students discuss or show the various ways they have modeled. Ensure that you highlight during discussions that pictures, objects, symbols and words are all models.

Ask, *Why do we model with math?* Have students discuss with a peer, then as a group. The goal is to recognize that modeling helps them solve problems.

Display a chart with Mathematical Practice 4. Restate Mathematical Practice 4 and have students assist in rewriting it as an "I can" statement, for example: **I can use pictures, objects, symbols, and words to solve math problems.** Have students draw examples of each model. Post the new "I can" statement and examples in the classroom.

Inquiry of the Essential Question:

How can I add and subtract two-digit numbers?

Inquiry Activity Target: **Students come to a conclusion that modeling helps solve two-digit computation.**

As an introduction to the chapter, present the Essential Question to students. The inquiry graphic organizer will offer opportunities for students to observe, make inferences, and apply prior knowledge of two-digit computation representing the Essential Question. As they investigate, encourage students to draw, write, and collaborate with peers to demonstrate their observations and thinking. Then have students present additional questions they may have to a peer to extend discussions.

Regroup students and restate Mathematical Practice 4 and the Essential Question. Pose questions to reflect on what has been learned to guide students in making connections between the Mathematical Practice and the Essential Question.

NAME _____ DATE _____

Chapter 6 Two-Digit Addition and Subtraction
Inquiry of the Essential Question:

How can I add and subtract two-digit numbers?

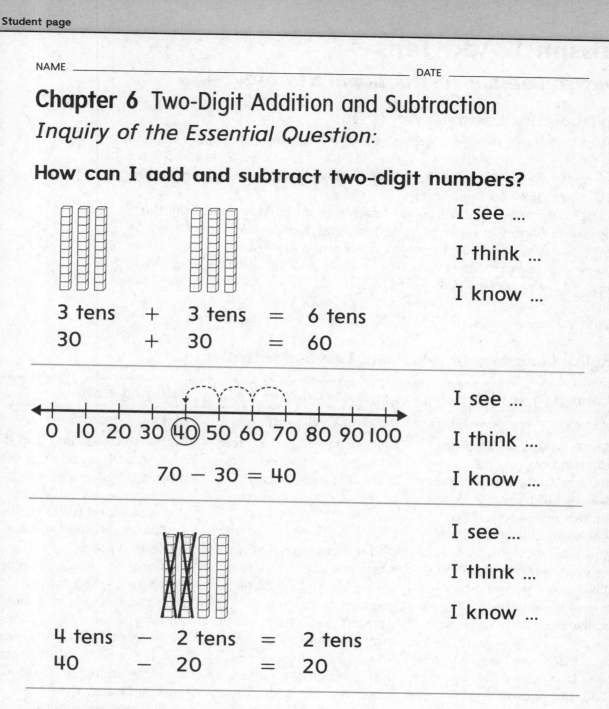

3 tens	+ 3 tens	=	6 tens
30	+ 30	=	60

I see ...

I think ...

I know ...

$$70 - 30 = 40$$

I see ...

I think ...

I know ...

4 tens	− 2 tens	=	2 tens
40	− 20	=	20

I see ...

I think ...

I know ...

Questions I have...

Teacher Directions: Read the Essential Question for students. Have students echo read. Direct students to describe their observations, inferences, and prior knowledge of each math example. Encourage students to write or draw additional questions they may have. Then have students share their ideas/questions with a peer.

Lesson 1 Add Tens

English Learner Instructional Strategy

Collaborative Learning: Act It Out

Select 5 students to form a group of 2 and a group of 3. Ask, *What is an addition number sentence for these groups of students?* **2 + 3 = 5** Ask, *How many students altogether?* **5** Write 2 + 3 = 5. Have students in the two groups hold up their fingers. *How many fingers are there in each group? Let's count by tens.* Have students count by tens to identify the number of fingers in each group. **2 tens and 3 tens**. Ask, *What is an addition number sentence using tens for these groups of students' fingers?* **2 tens + 3 tens = 5 tens** Ask, *How many tens altogether?* **5 tens** Write 2 tens + 3 tens = 5 tens. Ask, *How many in two tens?* **20** *How many in three tens?* **30** *How many in five tens?* **50** *How many fingers altogether?* **50** Write 20 + 30 = 50.

English Language Development Leveled Activities

Emerging Level	Expanding Level	Bridging Level
Activate Prior Knowledge Show groups of ten using different objects: ten fingers, thirty pencils, twenty students, ten books, and forty counters. Count each group and separate the total into groups of ten. Prompt students to name how many tens and how many in all with the following questions: *How many tens? How many in all?* Distribute base-ten rods and unit cubes. Demonstrate counting individual unit cubes to ten. Then line them up next to a tens rod. Ask, *How many tens?* **1** *How many in all?* **10**	**Building Oral Language** Demonstrate counting tens using groups of 10 counters. Show three groups of 10 counters. Count and then say, *Here are three tens. That's thirty counters.* Have students repeat after you when you say, *Three tens equals thirty.* Distribute counters to pairs of students. Call out a number that is a multiple of ten. Have students count out that number of counters, group the counters by tens, and describe the number using the sentence frame: _____ **tens equals** _____ .	**Making Connections** Demonstrate adding 2 and 4 using unit cubes. Push the groups together and say, *There are six in all.* Write 2 + 4 = 6 and have students read the sentence aloud in unison. Demonstrate adding 2 tens and 4 tens using ten rods. Push the ten rods together and say, *There are six tens in all.* Write 2 tens + 4 tens = 6 tens and have students read the sentence aloud in unison. Discuss the similarities between each problem. Repeat the activity, using 1 + 3 and 1 ten + 3 tens.

Teacher Notes:

NAME _____ DATE _____

Lesson 1 Number and Word Identification
Add Tens

Fill in the missing numbers and words. Then practice counting by tens.

Number	Tens	Word
20	__2__ tens	twenty
__30__	3 tens	thirty
__40__	4 tens	forty
10	__1__ ten	ten
50	__5__ tens	fifty
70	__7__ tens	seventy
__90__	9 tens	ninety
80	__8__ tens	eighty
__60__	6 tens	sixty

Teacher Directions: Use manipulatives to model each number. Model and then practice counting by tens as a group and then individually. For each item, have students identify the number or word. Then have students fill in the missing number or word in the other two columns. Encourage partners to report back about a number using a sentence frame such as: [Ninety] is [nine] tens.

Lesson 2 Count On Tens and Ones
English Learner Instructional Strategy

Collaborative Support: Partners Work

Have students work in pairs during the Explore and Explain Activity. Before you begin the activity model how to *count on* with a volunteer. Count to the number 10 and then have the student *count on* 3 more.

Distribute base-ten blocks to pairs of students. Say, *There are thirty-five people at a baseball game.* Have pairs count out 3 tens rods and 5 unit cubes. Then say, *Three more people come to the game.* Have one student count out 3 units cubes and hold them. Ask, *How many people are at the game in all? We will* **count on** *to find out.* Have the student holding the 3 unit cubes, place the cubes, one by one, into the group of 5. Have the other student count on, starting with **35, 35, 36, 37, 38** Ask, *How many in all?* **38** Write the number sentence, 35 + 3 = 38.

English Language Development Leveled Activities

Emerging Level	Expanding Level	Bridging Level
Word Knowledge Show ten connecting cubes. Say, *I will* **count** *these cubes.* Emphasize the word *count*, then count the cubes. Say, *Now I will* **count on** *from 4.* Demonstrate counting on starting at *4; 4, 5, 6, 7, 8, 9, 10* Distribute connecting cubes. Alternate between having students count all the cubes, and counting on from a number you provide. Before they count, have them say, **I will count.** Before they count on, have them say, **I will count on from ____.**	**Number Sense** Demonstrate the difference between counting and counting on using tens rods. Show 7 tens rods. Say, *I will* **count** *these tens.* Count the tens. Say, *I will* **count on** *from 4 tens.* Demonstrate counting on starting at *4 tens, 4 tens, 5 tens, 6 tens, 7 tens.* Distribute tens rods randomly to pairs. Have pairs count and identify how many tens using the sentence frame: **I counted ____ tens.** Group pairs and have them count on from one student's number of tens to find the pair's total number.	**Exploring Language Structure** Demonstrate counting unit cubes and use indicative forms of count. Say, *I will* **count** *ten unit cubes. I am* **counting** *ten unit cubes. I* **counted** *ten unit cubes.* Then demonstrate counting on from a given number. *I will* **count on** *from ____.* *I am* **counting on** *from ____.* *I* **counted on** *from ____.* Distribute connecting cubes. Have pairs of students count and count on while describing what they are doing using the terms: *count, counting, counted, count on, counting on* and *counted on.*

Teacher Notes:

NAME _____ DATE _____

Lesson 2 Concept Web
Count On Tens and Ones

Use the word web to complete examples of counting on by tens and ones.

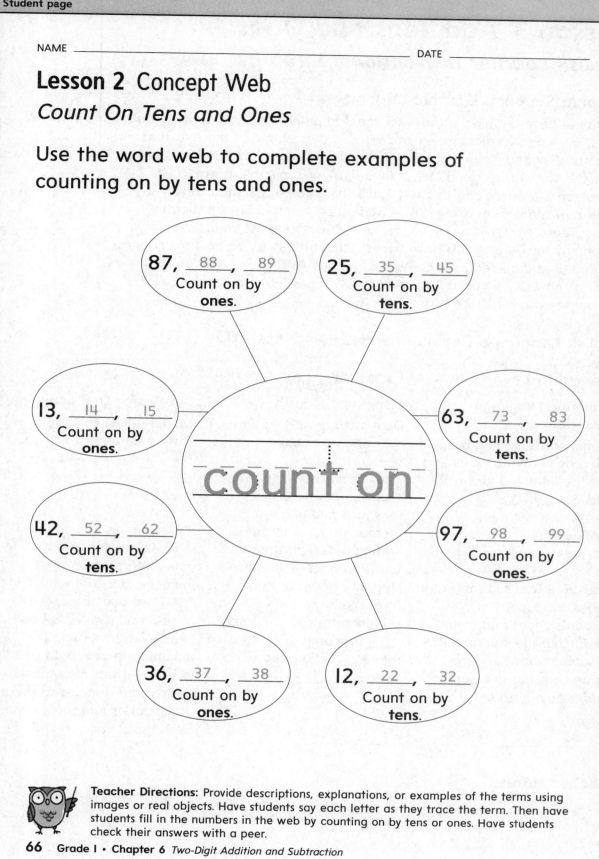

87, __88__, __89__
Count on by **ones**.

25, __35__, __45__
Count on by **tens**.

13, __14__, __15__
Count on by **ones**.

63, __73__, __83__
Count on by **tens**.

count on

42, __52__, __62__
Count on by **tens**.

97, __98__, __99__
Count on by **ones**.

36, __37__, __38__
Count on by **ones**.

12, __22__, __32__
Count on by **tens**.

Teacher Directions: Provide descriptions, explanations, or examples of the terms using images or real objects. Have students say each letter as they trace the term. Then have students fill in the numbers in the web by counting on by tens or ones. Have students check their answers with a peer.

Lesson 3 Add Tens and Ones
English Learner Instructional Strategy

Graphic Support: Graphic Organizers

Distribute base-ten blocks and work mats to pairs. Assign students to be "tens" or "ones." Direct students to work in the column of the mat that they are assigned. Say, *There are twenty-four children in a swimming pool.* Have the student assigned "tens" say aloud how many tens are in 24, **2** and place 2 tens rods on the mat. Have the student assigned "ones" say aloud how many ones are in 24, **4** and place 4 unit cubes on the mat. Say, *Three more children get in the pool.* Have "tens" students say aloud how many tens are in 3. **0** Have "ones" students say aloud how many ones are in 3, **3** and place 3 unit cubes on the mat. Ask, *Did we add to the tens column?* **no** *Did we add to the ones column?* **yes** Ask, *How many children are in the pool in all?* **27** Write 24 + 3 = 27.

English Language Development Leveled Activities

Emerging Level	Expanding Level	Bridging Level
Exploring Language Structure Demonstrate adding base-ten rods and unit cubes and use the inflected forms of add. Say, *I will **add** two tens and three ones. I am **adding** these rods and unit cubes. I **added** the numbers and got twenty-three.* Distribute tens rods and unit cubes. Have pairs of students count and count on. Encourage students to describe what they are doing by using the words: *add, adding, added.*	**Listen and Identify** Demonstrate adding tens rods and unit cubes. Say, *I will **add** two tens and three ones to make twenty-three.* Emphasize **add**. Distribute tens rods and unit cubes. Prompt students to model adding different groups of tens and ones. Before each student creates a model, have them use the following sentence frame: **I will add ____ tens and ____ ones to make ____.** Repeat as needed.	**Developing Oral Language** Model telling a number story such as, *Alice has four pieces of paper. Alex has twelve pieces of paper. How many piece of paper do they have altogether?* Demonstrate adding four ones to one ten and two ones. Write any two numbers (for example, 2 and 15). Have pairs of students create and solve addition number stories using the numbers. Repeat with additional sets of numbers. Encourage students to share their stories and the answers with the class.

Teacher Notes:

NAME _____ DATE _____

Lesson 3 Note Taking
Add Tens and Ones

Read the question. Write words you need help with. Use the lesson to write your Cornell notes.

Building on the Essential Question	Notes:

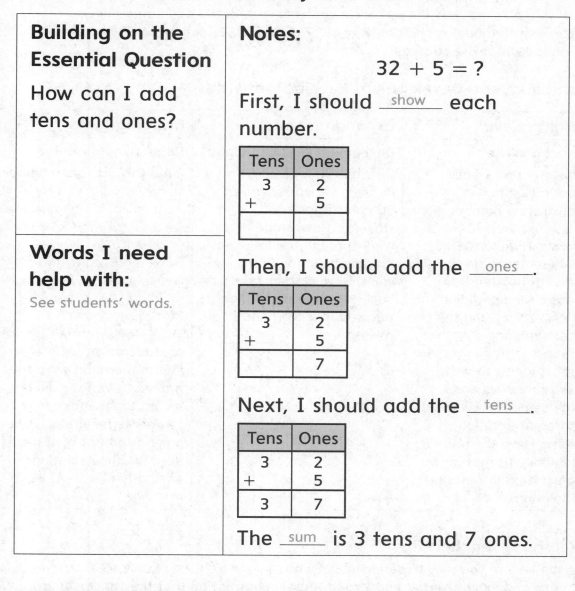

Building on the Essential Question

How can I add tens and ones?

Notes:

$$32 + 5 = ?$$

First, I should __show__ each number.

Tens	Ones
3	2
+	5

Then, I should add the __ones__.

Tens	Ones
3	2
+	5
	7

Next, I should add the __tens__.

Tens	Ones
3	2
+	5
3	7

The __sum__ is 3 tens and 7 ones.

Words I need help with:

See students' words.

Teacher Directions: Read the Building on the Essential Question and have students list words/phrases they need assistance with. Provide descriptions, explanations, or examples of the terms using images or real objects. Read each sentence frame and have students write the appropriate terms. Have students read their notes aloud.

Grade 1 • **Chapter 6** *Two-Digit Addition and Subtraction* **67**

Lesson 4 Problem Solving
Strategy: Guess, Check, and Revise
English Learner Instructional Strategy

Sensory Support: Visual Models

Distribute write-on/wipe-off boards to each student. Create "bowling balls" by cutting out circles from red, blue, and pink paper. Display 7 red circles, 21 blue circles, and 6 pink circles. Say, *Taye sees two colors of bowling balls. He sees 27 bowling balls in all. Which two colors does he see?* Have students write the 3 addition number sentences as you model each. $21 + 7 = 28$, $7 + 6 = 13$, $21 + 6 = 27$ Discuss the guess, check and revise strategy.

English Language Development Leveled Activities

Emerging Level	Expanding Level	Bridging Level
Listen and Write Use base-ten rods and number cubes to demonstrate a number story such as *I have two tens and four ones.* Then write the number sentence on the board with two tens and four ones. Model adding the numbers and writing the number sentence. *Two tens and four ones is twenty four.* Prompt students to write the tens and ones and number sentences as you say different addition problems. Have them use the sentence frames in your demonstration to describe their solutions.	**Developing Oral Language** Distribute the Lesson 4 Reteach (1) worksheet and 20 coins to pairs of students. Have students work through the four-step plan modeling with the coins. Have students report back to you using the terms: *guessed, checked* and *revised.*	**Deductive Reasoning** Read this addition number story. *Evan has three games in one box, six games in a second box, and four games in a third box. He picks two boxes and counts the total games in those boxes. He counted seven games. Which boxes did he choose?* Have small groups of students record the information and solve the problem. Groups should record the number sentences they used. Have pairs or groups of students describe their solutions to each other.

Multicultural Teacher Tip

During the lesson, you may experience ELs who appear to listen closely to your instructions and exhibit verbal and/or nonverbal confirmation that they understand the concepts. However, it becomes clear during the See and Show or On My Own parts of the lesson that the students did not actually understand. This may be due to a student coming from a culture in which the teacher is regarded as a strong, perhaps even intimidating, authority figure. They may be reluctant to ask questions, considering it impolite to do so and an implication that the teacher is failing.

NAME _____ DATE _____

Lesson 4 Problem Solving
STRATEGY: Guess, Check, and Revise

<u>Underline</u> what you know. (Circle) what you need
to find. Solve the problem.

picture

1. <u>**Harper** draws **12** pictures.</u>

   <u>**Raven** draws **some** pictures.</u>

   <u>Together **they** (Harper and Raven)
   drew **15** pictures.</u>

   (How many pictures **did Raven
   draw**?)

Part	Part
Whole	

<u>12</u> (+) <u>3</u> (=) <u>15</u>

Raven drew <u>3</u> pictures.

 Teacher Directions: Provide a description, explanation, or example of the boldface
terms and nouns using images or real objects. Read each sentence and have students
echo read. Encourage students to use the part-part-whole mat to help them visualize
what they know and need to know. Then have them write their answer in the restated
question. Have students read the answer sentence aloud.

Lesson 5 Add Tens and Ones with Regrouping

English Learner Instructional Strategy

Sensory Support: Manipulatives

Distribute base-ten blocks and work mats to pairs. Assign students to be "tens" or "ones." Students work in the column of the mat that they are assigned. Say, *Abby sees nineteen people flying kites.* Have each student say aloud the number of tens or ones in 19 and place tens rods or unit cubes on the mat, based on their assigned column. Say, *Abby sees three more people flying kites.* Have each student say aloud the number of tens or ones in 3 and place tens rods or unit cubes on the mat, based on their assigned column. Ask, *How many tens?* **1** *How many ones?* **12** *Have students trade 10 unit cubes for 1 tens rod.* **Ask,** *How many tens?* **2** *How many ones?* **2** Write 19 + 3 = 22.

English Language Development Leveled Activities

Emerging Level	Expanding Level	Bridging Level
Word Knowledge	**Exploring Language Structure**	**Act It Out**
Show ungrouped counters of two colors. Group them by color. Say, *This is a red group. This is a yellow group.* Stress the /g/ sound. Distribute 14 unit cubes to students. Have them count 10 cubes and place them in a group. Then have students put the 4 left over cubes in another group. Say, *Regroup the 10 ones for 1 ten.* Give each student a tens rod and have them give you 10 unit cubes to regroup. Prompt students to say, **regroup.** Say, *You have one ten and 4 ones. How many cubes do you have?* **14** Distribute 7 unit cubes to each student. Have them regroup to add 14 and 7.	Display a word web with regroup written in the center. Underline the prefix *re-* and say, *This part of the word means "again." When* **re-** *is added to the beginning of a word, it changes the word's meaning.* **Regroup** *means "group again."* Have students brainstorm other words that use the prefix *re-* and record them in the word web such as, *reread, redo, review, refreeze, reinjure,* etc. Have volunteers use the words in sentences to demonstrate meaning.	Show 13 individual connecting cubes. Say, *I will regroup to make a group of ten.* Create a train of 10 cubes and 3 left over. Model counting on from 10 to 13 as you point to the cubes. Distribute connecting cubes (greater than 10 but less than 20) to pairs. Have one partner group the cubes into sets of ten with some left over. The other partner counts on from 10 to tell how many. Have students describe the process using the words *group* and *regroup.*

Teacher Notes:

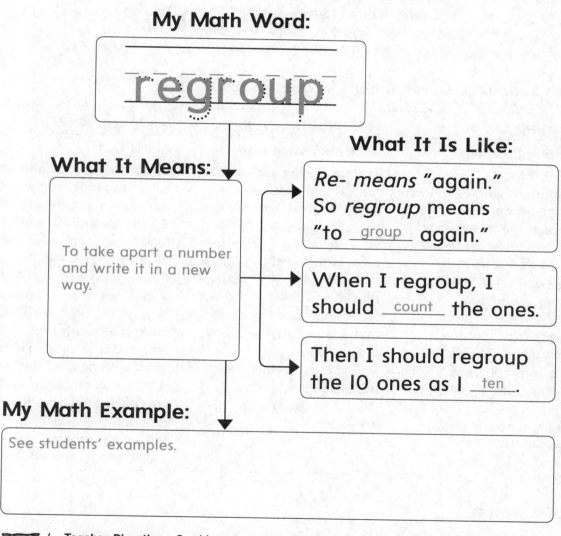

NAME _____ DATE _____

Lesson 5 Vocabulary Definition Map
Add Tens and Ones with Regrouping

Use the definition map to write what the math
word means and tell what the word is like.
Write or draw a math example.

My Math Word:

regroup

What It Means:

To take apart a number
and write it in a new
way.

What It Is Like:

Re- means "again."
So *regroup* means
"to __group__ again."

When I regroup, I
should __count__ the ones.

Then I should regroup
the 10 ones as I __ten__.

My Math Example:

See students' examples.

Teacher Directions: Provide a description, explanation, or example of the term using
images or real objects. Have students use the lesson or Glossary to define the math term.
Direct students to list characteristics, and draw a picture representing their math term.
Then encourage students to describe their picture to a peer.

Grade 1 · **Chapter 6** *Two-Digit Addition and Subtraction* **69**

Lesson 6 Subtract Tens

English Learner Instructional Strategy

Collaborative Learning: Act It Out

Form a line with 5 students. Count the students. Remove two students. Count again. Ask, *What is a subtraction number sentence for the remaining students?* **5 − 2 = 3** Ask, *How many students remain?* **3** Write 5 − 2 = 3. Repeat, but have students hold up their fingers. How many fingers are there with the five students? Let's count by tens. Have students count by tens to identify the number of fingers. **5 tens** Remove two students. Ask, *What is a subtraction number sentence using tens for the remaining students' fingers?* **5 tens − 2 tens = 3 tens** Ask, *How many tens of fingers remain?* **3 tens** Write 5 tens − 2 tens = 3 tens. Ask, *How many in five tens?* **50** *How many in two tens?* **20** *How many in three tens?* **30** *How many fingers remain?* **30** Write 50 − 20 = 30.

English Language Development Leveled Activities

Emerging Level	Expanding Level	Bridging Level
Exploring Language Structure	**Academic Vocabulary**	**Act It Out**
Distribute 8 pennies to each student. Model 8 − 5 = 3. Prompt students to model the subtraction sentence and say, **8 subtract 5 equals 3.** Then distribute 8 dimes and model 80 − 50 = 30. Prompt students to model the subtraction sentence and say, **8 tens subtract 5 tens equals 3 tens.** Then say, *80 subtract 50 equals 30.* Have students repeat chorally. Repeat with 7 − 1 = 6; 70 − 10 = 60.	Explain that when we are subtracting tens, we can hide the zeros and use the basic fact to solve. Write 50 − 30 = ____. Hide the zeros with self-stick notes to make 5 − 3 = ____. Explain that if 5 − 3 = 2, then we can bring back the zeros to make, 50 − 30 = 20. Say, *5 tens subtract 3 tens equals 2 tens.* Have students repeat chorally. Have students repeat with 70 − 20 = ____.	Use base-ten rods and unit cubes to model similar subtraction problems such as 7 − 6 = 1 and 70 − 60 = 10. Say, *I will subtract 6 from 7. Seven minus six equals one. I will subtract 6 tens from 7 tens. Seven tens minus six tens equals one ten.* Give pairs of students base-ten rods and unit cubes. Provide them with pairs of similar subtraction problems. As you visit each pair, have them describe their work, using the words: *subtract, take away, minus,* and *equals.*

Teacher Notes:

NAME _____ DATE _____

Lesson 6 Number Identification
Subtract Tens

Fill in the missing numbers. Then count back from 100 to 10.

1	2	3	4	5	6	7	8	9	
11	12	13	14	15	16	17	18	19	
21	22	23	24	25	26	27	28	29	
31	32	33	34	35	36	37	38	39	
41	42	43	44	45	46	47	48	49	
51	52	53	54	55	56	57	58	59	
61	62	63	64	65	66	67	68	69	
71	72	73	74	75	76	77	78	79	
81	82	83	84	85	86	87	88	89	
91	92	93	94	95	96	97	98	99	100

Teacher Directions: Use manipulatives to model each number. Model and then practice counting by tens as a group and then individually. Have students identify each missing number and write it in the blank. Count back from 100 to 10 by tens as a group. Encourage students to practice counting back by tens with a partner.

Lesson 7 Count Back by 10s
English Learner Instructional Strategy

Sensory Support: Diagrams

Draw or display a large 0–100 number line divided in increments of 10. Say, *Let's start on the number seventy. How many tens are in seventy?* **7** Have students help you identify the number 70 on the number line. Keep your finger by the number 70 and say, *I am pointing to seven tens and will count **back** by three tens.* Point to each number as you count back, saying, *sixty, fifty, forty.* Keep your finger by the number 40 and ask, *Now what number am I pointing to?* **40** Write 70 − 30 = 40. Repeat modeling how to count back by tens, having student volunteers point to numbers on the number line as the class chorally counts back. Write the words count back and say, *When you **count back** you are subtracting, or taking away.*

English Language Development Leveled Activities

Emerging Level	Expanding Level	Bridging Level
Word Knowledge	**Show What You Know**	**Making Connections**
Draw or display a number line from 1–100 in increments of ten. Say, *I will **count by** tens.* Model counting to 50 by tens. Say, *I will **count on** by tens.* Model counting on by tens to 80. Say, *I will **count back** by tens.* Model counting back by tens from 80 to 50. Have volunteers come to the number line and repeat what you have modeled. Give prompts that include *count by, count on, or count back by tens* using the number line.	Draw or display a number line from 1–100 in increments of ten. Demonstrate and explain the differences among *counting by, counting on,* and *counting back.* Have students copy the number line in their math journals. Play a game in which you prompt students to count by tens, count on, and count back and have students give the answer. For example: *Count on two tens from forty.* **60** After each prompt have students describe what they did using the terms: *count by, count on,* and *count back.*	Draw or display a number line from 1–100 in increments of ten. Use the number line to demonstrate the difference between the use of count in the terms *counting by, counting on,* and *counting back.* Have pairs of students discuss how to describe the differences between *counting by, counting on,* and *counting back.* Then have them share their conclusions with the class.

Teacher Notes:

NAME _____ DATE _____

Lesson 7 Note Taking
Count Back by 10s

Read the question. Write words you need help with. Use your lesson to write your Cornell notes.

Building on the Essential Question	**Notes:**
How can I count back by 10s?	**Word Bank** number difference count back subtract $$60 - 40 = ?$$ I can use a number line to ___subtract___ numbers by tens. ◄─┼──┼──┼──┼──┼──┼──┼──┼──┼──┼──► 0 10 20 30 40 50 60 70 80 90 100 I should start with the greater ___number___. ◄─┼──┼──┼──┼──┼──┼──●──┼──┼──┼──► 0 10 20 30 40 50 60 70 80 90 100 Then I should ___count___ ___back___ by tens. ◄─┼──┼──⌢─⌢─⌢──┼──●──┼──┼──┼──► 0 10 20 30 40 50 60 70 80 90 100 Where I stop is the ___difference___. So, $60 - 40 =$ ___20___.
Words I need help with: See students' words.	

Teacher Directions: Read the Building on the Essential Question and have students list words/phrases they need assistance with. Provide descriptions, explanations, or examples of the terms using images or real objects. Read each sentence frame and have students write the appropriate terms. Have students read their notes aloud.

Grade 1 • **Chapter 6** *Two-Digit Addition and Subtraction* **71**

Lesson 8 Relate Addition and Subtraction of Tens

English Learner Instructional Strategy

Vocabulary Support: Activate Prior Knowledge

Write the term related facts on the board. Say, *Related facts are a group of facts that use the same three numbers*. Have students call out two numbers that are less than ten. Explain that we will be using tens of these numbers. For example, if the number said was 2, then you will use 2 tens or 20. Write the numbers on large cards. Create the addition number sentence for the sum of those numbers and write the sum on another large card. For example, 20 + 30 = 50. Say, *Using these numbers I will create related subtraction number sentences*. Rearrange the cards to create 50 − 30 = 20 and 50 − 20 = 30.

As an extension, provide pairs of students with 3 tens that are related facts. Have them create three index cards with the related facts, one card with a minus sign, one card with an addition sign and one card with an equals sign. Students then model making related facts.

English Language Development Leveled Activities

Emerging Level	Expanding Level	Bridging Level
Word Knowledge Demonstrate related facts with base-ten blocks by showing and saying, *Eight minus five equals three. Three plus five equals eight.* Then write 8 − 5 = 3 and 3 + 5 = 8. Point to each number sentence and say, *fact*. Then point to both and say, *related facts*. Write another set of related facts and read the number sentences. Prompt students to identify each as a *fact* and both as *related facts*.	**Making Connections** Draw or show a picture of a family with grandparents, parents, and children. Explain that family members are *relatives* because they are *related*. They have a family *relationship*. Use connecting cubes to demonstrate related facts such as 70 + 20 = 90 and 90 − 20 = 70. Identify these as *related facts*. Have students describe the relationship between these facts. Have pairs of students write three pairs of related facts. Then encourage them to explain the relationships to a peer.	**Listen and Write** Use base-ten blocks to demonstrate a pair of related facts showing 30 + 40 = 70 and 70 − 30 = 40. Explain that each addition number sentence is a fact and that these are related facts because they use the same three numbers. Discuss with students how the numbers are related. Have students identify and write related facts using these numbers: 50, 40, and 90.

Teacher Notes:

NAME _____ DATE _____

Lesson 8 Four-Square Vocabulary
Relate Addition and Subtraction of Tens

Trace the words. Write the definition for *related facts*. Write what the word means, draw a picture, and write your own sentence using the word.

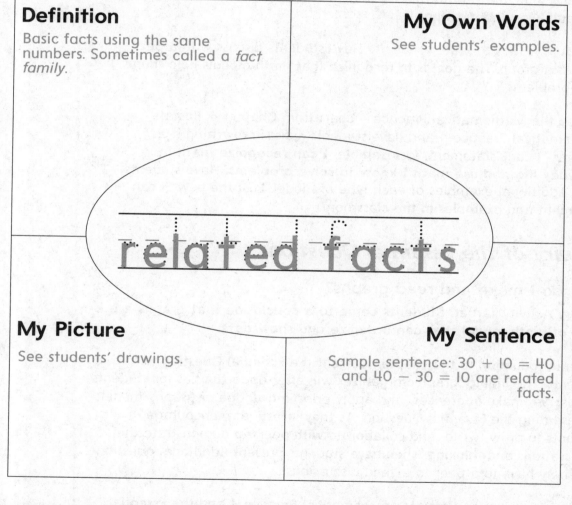

Definition

Basic facts using the same numbers. Sometimes called a *fact family*.

My Own Words

See students' examples.

related facts

My Picture

See students' drawings.

My Sentence

Sample sentence: $30 + 10 = 40$ and $40 - 30 = 10$ are related facts.

Teacher Directions: Provide a description, explanation, or example of the term using images or real objects. Have students use the Glossary to write the definition. Direct students to write a definition in their own words and draw a picture representing their math term. Have students write a sentence using the term and then encourage students to read their sentence to a peer.

72 Grade I · **Chapter 6** *Two-Digit Addition and Subtraction*

Chapter 7 Organize and Use Graphs

What's the Math in This Chapter?

Mathematical Practice 4: Model with mathematics

Display a large classroom 2-column chart to discuss the multiple meaning word *model*. Show images of models such as: fashion models, car and airplane models, sports role models, connecting cubes, number line, or student using a math manipulative, and so on. Have students sort the images into 2 groups on the 2-column chart for Math Meaning and Non-Math Meaning. Compare and contrast the two groups.

Ask, *Have you used math models?* Solicit **Yes** from students, then have students show the various models they have used thus far utilizing manipulatives and drawings.

Ask, *Why do we use math models?* Have students discuss with a peer, then as a group. The goal is to recognize that math models help them solve problems.

Display the Mathematical Practice 4 chart from Chapter 6. Restate Mathematical Practice 4 and have students assist in rewriting it as another "I can" statement, for example: **I can recognize math in everyday life and use math I know to solve problems.** Have students draw additional examples of each type of model. Post the new "I can" statement and examples in the classroom.

Inquiry of the Essential Question:

How do I make and read graphs?

Inquiry Activity Target: **Students come to a conclusion that graphs are a type of math model that can organize and show data.**

As an introduction to the chapter, present the Essential Question to students. The inquiry graphic organizer will offer opportunities for students to observe, make inferences, and apply prior knowledge of graphs/models representing the Essential Question. As they investigate, encourage students to draw, write, and collaborate with peers to demonstrate their observations and thinking. Then have students present additional questions they may have to a peer to extend discussions.

Regroup students and restate Mathematical Practice 4 and the Essential Question. Pose questions to reflect on what has been learned to guide students in making connections between the Mathematical Practice and the Essential Question.

NAME _____ DATE _____

Chapter 7 Organize and Use Graphs
Inquiry of the Essential Question:

How do I make and read graphs?

Favorite Ice Cream		
Ice Cream	Tally	Total
Chocolate	ЖН III	8
Vanilla	ЖН	5
Strawberry	III	3

I see ...

I think ...

I know ...

Favorite Season

	Fall						
	Spring						
	Summer						

I see ...

I think ...

I know ...

Favorite Ice Cream

Chocolate									
Vanilla									
Strawberry									

0 1 2 3 4 5 6 7 8 9

I see ...

I think ...

I know ...

Questions I have...

_____

— — — — — — — — — — — — — — — — —

_____

— — — — — — — — — — — — — — — — —

_____

Teacher Directions: Read the Essential Question for students. Have students echo read. Direct students to describe their observations, inferences, and prior knowledge of each math example. Encourage students to write or draw additional questions they may have. Then have students share their ideas/questions with a peer.

Lesson 1 Tally Charts
English Learner Instructional Strategy

Graphic Support: Tables and Graphs

What is your favorite pizza topping? Record each student's answer in a tally chart. Say, *Asking people the same question and collecting their answers is how to take a **survey**. I **surveyed** some of you about your favorite pizza topping.* Give an example of how the word *survey* can be used as a noun or a verb. Give an example of each use of the word.

Point to the tally marks on the chart and say, *For each answer, I put a mark on this table.* Survey another student and place a tally mark on the tally chart to record his or her answer. Say, *Each mark is called a **tally mark**. A table with **tally marks** is called a **tally chart**.*

English Language Development Leveled Activities

Emerging Level	Expanding Level	Bridging Level
Word Knowledge Draw a tally chart with the heading *Girls* and *Boys*. Say, *tally chart.* Count the number of boys and write a tally mark for each. As you write each *tally mark,* say, tally mark. Repeat this step for the number of girls. Use the following sentence frame to describe your groups' data: *The tally chart shows ____ boys and ____ girls.* Guide students to help you complete a tally chart to survey the students' different hair colors. Emphasize the words *tally chart* and *survey.*	**Academic Vocabulary** Say, *Let's take a **survey** and make a **tally chart** of hair colors.* Draw a tally chart and with hair color headings. Prompt students to count the number with each hair color as you make tally marks on the chart. Summarize the data, *The **tally chart** shows there are ____ students with ____ hair.* Have pairs create tally charts to show how many boys and girls in the class. Have them summarize using this sentence frame: **Our tally chart shows ____ boys and ____ girls.**	**Exploring Language Structure** Draw a tally chart then conduct a survey to record how many boys and girls. Say, *This is a **tally chart**. I am going to take a **survey**. I am **surveying** the number of boys and girls in the class. As I count the number of boys/girls I am **tallying** the number. My tally chart now has data. I **surveyed** the group and then I **tallied** the information.* Discuss the data. Have pairs of students conduct a survey of different colors of hair in the class and use a tally chart to record the data. Have students report their data using different tenses of tally and survey.

Multicultural Teacher Tip

Some ELs may write tally marks in a somewhat different format than the familiar four vertical lines with one diagonal. In Latin America, tally marks are generally drawn as four successive lines that form a square, with the fifth mark being a diagonal across the square. Students from Asia will use marks that eventually form the following character consisting of five marks: 正

NAME _____ DATE _____

Lesson I Vocabulary Word Study
Tally Charts

Circle the correct word to complete the sentence.

I. A survey asks people the same _____.

answer   (question)

Show what you know about the word:

survey

There are __6__ letters.

There are __2__ vowels.

There are __4__ consonants.

__2__ vowels + __4__ consonants = __6__ letters in all.

Draw a picture to show what the word means.

See students' examples.

 Teacher Directions: Provide a description, explanation, or example of the new term using images or real objects. Read the sentence and have students circle the correct word. Direct students to count the letters, vowels and consonants in the math term, then complete the addition number sentence. Guide students to draw a picture representing their math term. Then encourage students to describe their picture to a peer.

Lesson 2 Problem Solving Strategy: Make a Table

English Learner Instructional Strategy

Collaborative Support: Signal Words/Phrases

Create an anchor chart listing signal words and phrases for addition and subtraction using examples from the lesson such as: *how many more ... than, in all, altogether*. Have students work in pairs to identify and then highlight these signal words/phrases in the exercises.

Have pairs work together to first create a tally chart for Exercise 1 then transfer the data into the table. One student counts the number of each animal aloud while the second records tally marks in a table for each number. Then have pairs work together to solve Exercises 1–7.

English Language Development Leveled Activities

Emerging Level	Expanding Level	Bridging Level
Riddle Me This	**Word Knowledge**	**Show What You Know**
Draw a table showing data of how many girls and boys are in the class. Say, *This is a table*. Then contrast it with an example of a table you currently use in the classroom. Say, *This is also called a table*. Draw a data table and a furniture table. Prompt students to identify the correct type of table as you provide riddles such as: *I sit at this. I make this to help me solve a problem. This can show data. I can use this to help me when I eat.*	Identify and point to students' different shirt colors. Say each color as you point. Draw a table with columns for shirt color and number. Say, *This is a table*. Have students with red shirts raise their hands and say, **red shirt**. Then count the number and record it on the table. Repeat for students wearing brown and blue shirts. Point to each shirt color or number recorded in the table. Have students repeat the category name after you as you say it aloud.	Have groups of students conduct a short survey such as how many brothers and sisters each member of the group has. Use a table to represent the responses showing the number of students with 0, 1, 2, 3, or more than 3 siblings. Have each group present their table to the class using the words: *table, survey,* and *tally*.

Teacher Notes:

NAME _____ DATE _____

Lesson 2 Problem Solving
STRATEGY: *Make a Table*

<u>Underline</u> what you know. Circle what you need to find. Make a table.

I. **Ana's** toy has 4 wheels.

Corey's toy has I wheel.

Bryn's toy has 2 wheels.

The toys are a **unicycle,** a **bicycle,** and a **toy car.**

Who has the bicycle?

unicycle   toy car

bicycle

← wheel

Name	Wheels	Riding Toy
Ana	4	toy car
Bryn	2	bicycle
Corey	I	unicycle

Bryn has the bicycle.

 Teacher Directions: Provide a description, explanation, or example of the boldface terms and nouns using images or real objects. Read each sentence and have students echo read. Encourage students to use the table to organize the information and then write their answer in the restated question. Have students read the answer sentence aloud.

Grade I • **Chapter 7** *Organize and Use Graphs* **75**

Lesson 3 Make Picture Graphs

English Learner Instructional Strategy

Vocabulary Support: Utilize Resources

Before the lesson, direct students to review the Glossary definitions for *data, graph,* and *picture graph* in both English and Spanish. Utilize other appropriate translation tools for non-Spanish speaking ELs.

Ask each student the following question, *What is your favorite color?* Record each student's answer in a tally chart. Point to the tally marks on the tally chart and say, *For each answer, I put a tally mark. The answers collected are called the data.* Have students chorally say, **data**. *I can record the data in a tally chart or show the data in a graph.* Use the data to create a picture graph. Say, *For each answer, I can also put a picture. A graph that shows data with pictures is called a picture graph.* Have students chorally say, **picture graph**.

English Language Development Leveled Activities

Emerging Level	Expanding Level	Bridging Level
Word Recognition Conduct a survey of students' favorite fruits. Record the data in a tally chart. Point to the tally chart and ask, *What is this?* **tally chart** Circle the tally marks and ask, *What are these marks?* **tally marks** Write data on the board. Say, *This is data. The tally marks are data.* Point to the term and the data in the chart and have students repeat, **data**. Use the data to create a picture graph. Write picture graph. Point to the term and the picture graph and say, *picture graph.* Have students repeat.	**Show What You Know** Show a tally chart of the different colors of hair represented by 10 randomly selected students in the class. Describe the tally chart and summarize the data. Guide students to help you make a picture graph of the data using different colored circles to represent the data. Have them describe the picture graph and summarize the data. Ask, *What are the differences between the tally chart and the picture graph?* Lead them to determine that the data is the same but the representation is different.	**Exploring Language Structure** Display a picture graph. Say, *A picture graph uses pictures to show data.* Display a word web with *picture graph* written in the center. Underline the word *graph* and say, *To graph is to write or record data.* Have students brainstorm other words with the root word *graph* (photograph, autograph, graphite, calligraphy, geography). If possible display pictures to represent each word. Have volunteers use the words in sentences to demonstrate meaning.

Teacher Notes:

NAME _____ DATE _____

Lesson 3 Four-Square Vocabulary
Make Picture Graphs

Trace the words. Write the definition for *picture graph*. Write what the words mean, draw a picture, and write your own sentence using the words.

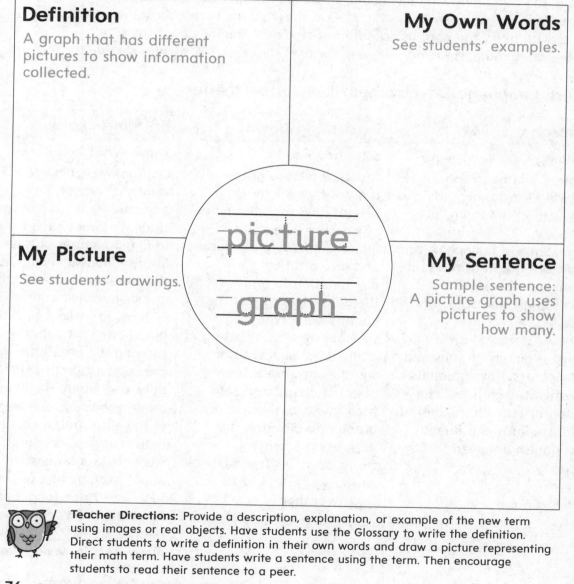

Definition
A graph that has different pictures to show information collected.

My Own Words
See students' examples.

My Picture
See students' drawings.

picture

graph

My Sentence
Sample sentence:
A picture graph uses pictures to show how many.

Teacher Directions: Provide a description, explanation, or example of the new term using images or real objects. Have students use the Glossary to write the definition. Direct students to write a definition in their own words and draw a picture representing their math term. Have students write a sentence using the term. Then encourage students to read their sentence to a peer.

Lesson 4 Read Picture Graphs

English Learner Instructional Strategy

Vocabulary Support: Modeled Talk

Display a picture graph, similar to the Explore and Explain picture graph. Use large cut out pictures of vegetables to represent the data.

Say, *This is a* **picture graph**. *The pictures represent* **data**. Describe each row of data using the sentence frame, *Each picture of a _____ stands for a person who chose _____ as his or her favorite vegetable.*

Have 4 students each remove 1 carrot picture from the graph and stand holding it. Have the remaining students count aloud the number of students holding carrots. Say, *Four people like carrots.* Have those students remain standing and repeat with peas and corn using different students. Finally, count the total number of students standing and say, *Thirteen people were* **surveyed** *in all.*

English Language Development Leveled Activities

Emerging Level	Expanding Level	Bridging Level
Building Oral Language	**Sentence Frame**	**Show What You Know**
Draw a picture graph showing how many girls and boys are in the room. Say, *This is a picture graph of data.* Show a tally chart and a picture graph of the same data. Point to each accordingly and say, *This is a tally chart. This is a picture graph.* Show other pairs of picture graphs and tally charts. Have students identify if each is a picture graph or tally chart. Repeat until students can fluently distinguish between the types.	Create a picture graph of the weather for the past several days with the types of weather listed and pictures of a sun, a cloud, and rain or other precipitation to represent the data. Have students identify each type of weather and count how many days each type of weather occurred. Then direct students to create their own weather picture graph and describe the data using a sentence frame such as, **The data shows _____ days with _____ weather.**	Distribute scissors and a variety of grocery advertisements to groups of students. Direct each student to find 3 pictures of their favorite foods and cut out the pictures. Then have the groups use their cut out pictures to make a picture graph on chart paper showing the data from their group. Assist groups with titles and labels. Have groups present their picture graphs to the class and discuss the data collected. Encourage students to ask questions about each picture graph, such as: **Which food was liked the most? How many people chose _____? How many people were surveyed in all?**

Teacher Notes:

NAME _____ DATE _____

Lesson 4 Vocabulary Sentence Frames
Read Picture Graphs

The math words in the word bank are for the sentences below. Write the words that fit in each sentence on the blank lines.

Word Bank		
graph	data	picture graph

1. A _picture_ _graph_ uses pictures to show information.

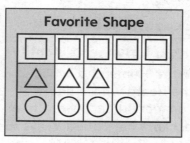

Favorite Shape

2. Information like: △ △ is called _data_ .

3. A tally chart is a type of _graph_ .

Favorite Shape

Shape	Tally	Total
□	||||	4
△	||	2
○	|||	3

Teacher Directions: Provide a description, explanation, or example of the each term using images or real objects. Read each sentence frame and have students echo read. Direct students to write the correct terms in each blank. Then encourage students to read each sentence to a peer.

Grade 1 • **Chapter 7** *Organize and Use Graphs* **77**

Lesson 5 Make Bar Graphs
English Learner Instructional Strategy

Vocabulary Support: Utilize Resources

Before the lesson, direct students to review the Glossary definitions for *data, graph,* and *bar graph* in both English and Spanish. Utilize other appropriate translation tools for non-Spanish speaking ELs.

Ask each student the following question, *What is your favorite color?* Record each student's answer in a tally chart. Point to the tally marks on the tally chart and say, *For each answer, I put a tally mark. The answers collected are called the* **data**. *I can record the* **data** *in a tally chart or show the* **data** *in a* **graph**. Use the data to create a bar graph. Say, *For each answer, I can also shade in a box. A* **graph** *that shows data using bars is called a* **bar graph**.

English Language Development Leveled Activities

Emerging Level	Expanding Level	Bridging Level
Word Recognition Conduct a survey of favorite fruits or vegetables. Record data in a tally chart. Say, *This is a* **tally chart**. Use the data to create a picture graph. Say, *This is a* **picture graph**. *Use the data to create a bar graph.* Say, *This is a* **bar graph**. Discuss the characteristics of each. Display other picture graphs, tally charts, and bar graphs. Have students identify which is which. Discuss that tally marks, pictures and bars all show data.	**Look, Listen, and Identify** Write Home and School on the board with a line connecting each word to a pocket chart titled Ways to Get to School. Create pictures of a car, bus, bicycle, and feet. Move each picture along the line and say, *I get to school by _____.* Have students draw a picture of how they get to school on index cards, place their picture in the pocket chart, and describe the mode of transport using the sentence frame. Introduce the term *bar graph* and create a bar graph from the data.	**Background Knowledge** Write *bar graph* on the board. Say, *bar graph,* and have students repeat. Show students an example of a bar graph and discuss its components. Post a tally chart and picture graph that contain the same data on Favorite Summer Activities. Ask, *Can you make a* **bar graph** *that shows this information?* Provide students with a blank table to graph the data. Ask students which type of graph is easier for them to create and read. Have students describe the data in their bar graph using complete sentences. For example, **Six students like swimming as their favorite summer activity.**

Teacher Notes:

NAME _____ DATE _____

Lesson 5 Vocabulary Definition Map
Make Bar Graphs

Use the definition map to write what the math word means and tell what the word is like. Write or draw a math example. Share your examples with a classmate.

My Math Word:

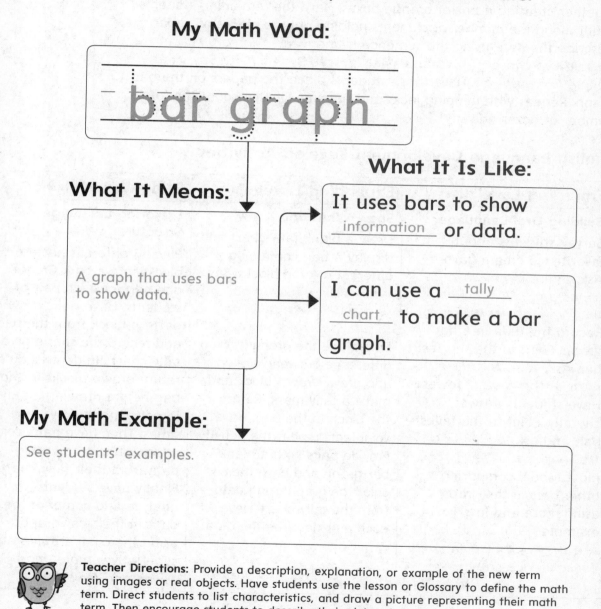

bar graph

What It Means:

A graph that uses bars to show data.

What It Is Like:

It uses bars to show ___information___ or data.

I can use a ___tally___ ___chart___ to make a bar graph.

My Math Example:

See students' examples.

Teacher Directions: Provide a description, explanation, or example of the new term using images or real objects. Have students use the lesson or Glossary to define the math term. Direct students to list characteristics, and draw a picture representing their math term. Then encourage students to describe their picture to a peer.

Lesson 6 Read Bar Graphs
English Learner Instructional Strategy

Vocabulary Support: Modeled Talk

Display a large, demonstration bar graph, with similar titles and headings to the Explore and Explain bar graph. Prepare square pieces of purple, green, and orange colored paper.

Say, *This is a **bar graph**. We fill in boxes to create bars that represent data.* Have 4 students each take 1 orange piece of paper and stand together creating a bar of orange boxes. Have the remaining students count aloud the number of students holding orange pieces of paper. Describe the data using the sentence frame: *Each colored box stands for a person who chose _____ as his or her favorite gym activity.* Say, *Four people like push-ups.* Have those students place the papers on the bar graph. Repeat with jumping jacks and running. Finally, count the total number of boxes and say, *Eleven people were surveyed in all.*

English Language Development Leveled Activities

Emerging Level	Expanding Level	Bridging Level
Building Oral Language Survey students about favorite ice cream flavors. Ask, *What is your favorite ice cream flavor, vanilla, chocolate or strawberry?* Record the data in a bar graph. Point to the shortest bar. Ask, *Does this show the most or the fewest?* **fewest** Have students answer chorally. Point to the tallest bar, and ask, *Does this show the most or the fewest?* **most** Model comparing other bars in the graph using more and less. For example, *Do more students like strawberry ice cream or chocolate ice cream?*	**Show What You Know** Draw a blank tally chart and display a bag containing 3 different colored blocks. Say, *I will **draw** a block from the bag and write a tally mark to show the color I **drew**.* Emphasize the pronunciation difference between *draw* and *drew*. Draw 1 block and make a tally mark. Return the block to the bag. Have 9 volunteers repeat the activity. Provide pairs with a blank bar graph and have them color the graph using data from the tally chart. Have each pair describe the data.	**Exploring Language Structures** Place 3 different pattern blocks into a bag. Create one bag for each pair of students. Have pairs draw a pattern block from the bag and record the shape on a tally chart. Students should return drawn blocks to the bag before drawing again. Direct pairs to draw ten times. Then have pairs color a blank bar graph using data from their tally chart. Finally have students analyze and compare the data in their bar graphs using the comparatives and superlatives *more, most, fewer,* and *fewest.*

Teacher Notes:

NAME _____ DATE _____

Lesson 6 Note Taking
Read Bar Graphs

Read the question. Write words you need help with.
Use your lesson to write your Cornell notes.

Building on the Essential Question	Notes:
Building on the Essential Question How can I read bar graphs?	**Notes:** I know the bars on a bar graph tell __how__ __many__. The __bars__ on a bar graph can be horizontal or vertical. I should look where each bar __ends__. Then I should read the __number__.

Words I need help with:

See students' words.

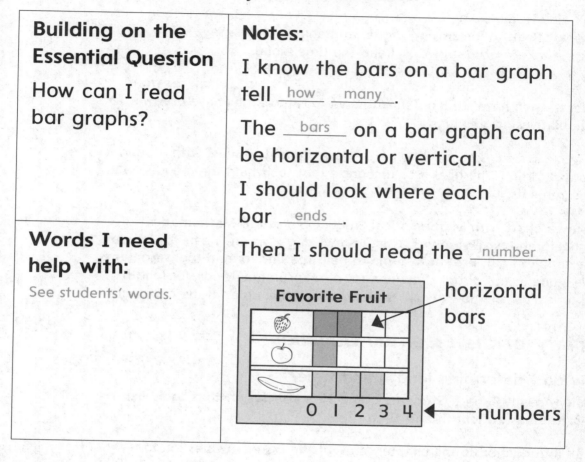

Favorite Fruit

horizontal bars

0 1 2 3 4 ← numbers

Teacher Directions: Read the Building on the Essential Question and have students list words/phrases they need assistance with. Provide descriptions, explanations, or examples of the terms using images or real objects. Read each sentence frame and have students write the appropriate terms. Have students read their notes aloud. Direct students to draw a picture representing the question. Then encourage students to describe their picture to a peer.

Grade 1 · Chapter 7 *Organize and Use Graphs* **79**

Chapter 8 Measurement and Time
What's the Math in This Chapter?

Mathematical Practice 5: Use appropriate tools strategically

Display a crayon and model finding the length of the crayon using the connecting cubes. Say, *These cubes are my **tool** to find the length of this crayon.* Read the length in cubes aloud. Have students practice measuring classroom objects using cubes.

Think aloud. Say, *I wonder what time it is... What tools can I use to find the time?* Point to, or display, digital and analog clocks. Say, *My tools to measure time are these clocks!* Read the time aloud.

Ask, *Have you ever used a tool to find length or time?* Solicit **Yes** from students, then have students show the various tools they have used via manipulatives or drawings.

Ask, *How can tools help us measure?* Have students discuss with a peer, then as a group. The goal is to recognize that tools help them understand measurements.

Display a chart with Mathematical Practice 5. Restate Mathematical Practice 5 and have students assist in rewriting it as an "I can" statement, for example: **I can use math tools to help me understand measurements.** Have students cut out or draw images of measurement tools for length and time. Post the new "I can" statement and examples in the classroom.

Inquiry of the Essential Question:

How do I determine length and time?

Inquiry Activity Target: **Students come to a conclusion that tools can be used to measure length and time.**

As an introduction to the chapter, present the Essential Question to students. The inquiry graphic organizer will offer opportunities for students to observe, make inferences, and apply prior knowledge of length and time representing the Essential Question. As they investigate, encourage students to draw, write, and collaborate with peers to demonstrate their observations and thinking. Then have students present additional questions they may have to a peer to extend discussions.

Regroup students and restate Mathematical Practice 5 and the Essential Question. Pose questions to reflect on what has been learned to guide students in making connections between the Mathematical Practice and the Essential Question.

NAME _____ DATE _____

Chapter 8 Measurement and Time
Inquiry of the Essential Question:

How do I determine length and time?

It's nine o'clock.

It's nine o'clock.

I see ...

I think ...

I know ...

The pencil is 6 connecting cubes long.

I see ...

I think ...

I know ...

Longest

Shortest

I see ...

I think ...

I know ...

Questions I have...

_____

- -

_____

- -

_____

Teacher Directions: Read the Essential Question for students. Have students echo read. Direct students to describe their observations, inferences, and prior knowledge of each math example. Encourage students to write or draw additional questions they may have. Then have students share their ideas/questions with a peer.

Lesson 1 Compare Lengths

English Learner Instructional Strategy

Sensory Support: Realia

Gather three different lengths of boxes like those in the Explore and Explain activity. Label them: Box A, Box B, and Box C. Write the terms *shorter than* and *longer than* on the board. Discuss the length of each box in comparison to the others as you gesture to the appropriate written terms. Have students say comparisons such as: **Box A is shorter than box C. Box B is longer than box A. Box C is longer than Box A and Box B.**

English Language Development Leveled Activities

Emerging Level	Expanding Level	Bridging Level
Academic Vocabulary	**Word Knowledge**	**Act It Out**
Find an item that is 3 connecting cubes in length. Model and display a train of 3 connecting cubes horizontally next to the item and say, *These connecting cubes show length. This ____ (item name) is three connecting cubes in length.* Emphasize the /l/ sound. Repeat with another item and 10 cubes. Compare the cube trains. Point to the shorter train and say, *This cube train is shorter.* Point to the longer cube train and say, *This cube train is longer.* Have students repeat chorally.	Connect 3 connecting cubes and say, *These show length.* Repeat with ten cubes. Compare the lengths. Say, *This cube train is **shorter than** this cube train. This cube train is **longer than** this cube train.* Emphasize shorter than and *longer than*. Have students find 3 items of different lengths and lay the items on their desks to compare lengths using the following sentence frames: **The ____ (item name) is longer than the ____ (item name). The ____ (item name) is shorter than the ____ (item name).**	Demonstrate length by having three volunteers lay on the floor in positions that are parallel to each other. Ask, *Is (student name) longer than or shorter than (student name)?* Have students respond in a complete sentence using one of the following sentence frames:**____ (student name) is longer than ____ (student name). ____ (student name) is shorter than ____ (student name).** Repeat until all students have had a chance to participate.

Multicultural Teacher Tip

In a multicultural classroom, there may be differences in students' perceptions of appropriate physical space or physical contact. For example, a student from Japan might be offended by an encouraging pat on the shoulder, as casual physical contact is generally avoided in Japanese culture. However, a student from Latin America would consider it a friendly and common gesture. Americans, on average, consider a 2-foot boundary as an appropriate amount of physical space, but in Latin American, Middle Eastern, and European cultures, the boundary is smaller, and casual touching during a conversation is not seen as inappropriate.

NAME _____   DATE _____

Lesson I Word Identification
Compare Lengths

Match.

short

long

length

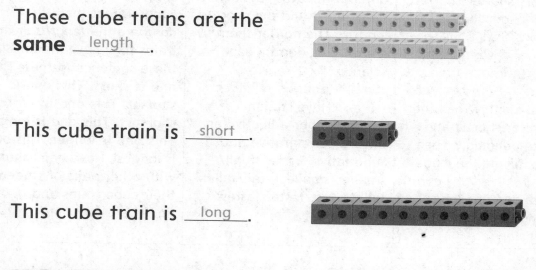

Write the correct word from above for each sentence on the blank lines.

These cube trains are the
same ___length___ .

This cube train is ___short___ .

This cube train is ___long___ .

Teacher Directions: Review the words using images or real objects. Have students say each word and then draw a line to match the word to its meaning. Direct students to say each word and then write the corresponding words in the sentence frames. Encourage students to read the sentences to a peer.

Grade I • **Chapter 8** *Measurement and Time*   **81**

Lesson 2 Compare and Order Lengths
English Learner Instructional Strategy

Vocabulary Support: Modeled Talk

Display an object, such as a paper clip. Say, *I will find an object that is longer than this one*. Display an object that is longer, such as a pencil. Line the objects up to compare length. Point to the part of the new object that extends past the original object and say, *The is longer than the ____*. Point to the pencil. Say, *I will find an object that is longer*. Display an object that is longer, such as a paintbrush. Line the objects up to compare length. Point to the part of the new object that extends past the previous object and say, *The ____ is longer than the ____*. Hold up and describe each of the three objects as long, longer, and longest. Repeat this activity with different classroom objects emphasizing the terms *short, shorter* and *shortest*.

English Language Development Leveled Activities

Emerging Level	Expanding Level	Bridging Level
Academic Vocabulary	**Exploring Language Structure**	**Word Knowledge**
Show lengths of 3, 5, and 10 connecting cube trains. Run your finger along the horizontal length of each and say, *These show length*. Lay each down in order from longest to shortest and model short/shorter/shortest and long/ longer/longest comparisons. Say, *I will compare the lengths*. Point to the shortest cube train. Say, *This is the shortest cube train*. Then identify the other cube trains as shorter and short correspondingly. Then model identifying the longest. Say, *This is the longest cube train*. Then describe long/ longer/longest, using the preceding steps.	Explain that the *-er* ending is used to compare two things, and the *-est* ending compares three or more things. Show 3 trains of connecting cubes that are different lengths. Identify the longest and the shortest lengths. Then order them by length and identify each saying, *short, shorter, shortest* and *long, longer, longest*. Have students suggest other adjective or adverb comparative and superlative forms: small/ smaller/smallest, tall/taller/ tallest, fast/faster/fastest.	Model the words *long, longer, longest* and *short, shorter, shortest* using different lengths of connecting cube trains. Distribute connecting cubes to student pairs. Have pairs create 3 different lengths of cubes. Direct them to *compare* the *lengths* of the trains they have made using these sentence frames: **This one is short. This one is shorter. This one is shortest. This one is long. This one is longer.** This one is longest. Have pairs share with other pairs of students their cube trains and findings.

Teacher Notes:

NAME _____ DATE _____

Lesson 2 Concept Web
Compare and Order Lengths

Look at the circled picture. Then circle the correct word.

 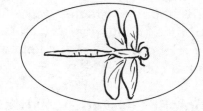 

long/longer/longest

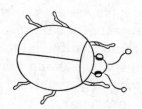 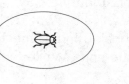

short/shorter/shortest

 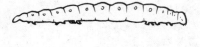 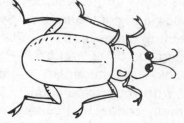

short/shorter/shortest

 Teacher Directions: Provide a description, explanation, or example of the new term using images or real objects. Direct students to look at the circled picture and then choose the correct word to describe it. Model and have students practice sentences that compare using sentence frames such as: **The butterfly is longer than the bee. The dragonfly is the longest.**

Lesson 3 Nonstandard Units of Length
English Learner Instructional Strategy

Sensory Support: Manipulatives

Display classroom objects of different lengths. Say, *We will measure the length of each object using connecting cubes.* Model how to measure one of the objects. Explain the importance of lining up the cubes with the edge of the object. Say, *I connect the cubes until I reach the other end of the object. Then I will count to find how many cubes.* Have students count the number of cubes aloud and say the length using the sentence frame: **The _____ is about _____ cubes long.** Write the measurement using the words about _____ cubes. Say, *We use the word about because the measurement is not an exact number of cubes. The _____ is not exactly _____ cubes, so we say it is about _____ cubes.* Repeat, having volunteers measure the other objects with the cubes.

English Language Development Leveled Activities

Emerging Level	Expanding Level	Bridging Level
Word Knowledge Ask a volunteer to measure a book's length using connecting cubes. Say, *Please measure the book. Use the connecting cubes to measure.* Emphasize the word *measure*. Discuss how to measure properly, lining up the cubes at one end of the book and continue connecting cubes until reaching the other end of the book. Have students count with you as you count how many connecting cubes long the book is. Say, *The book is about _____ cubes in length.* Encourage students to repeat chorally.	**Exploring Language Structure** Say, *The word measure describes what you do to find the length, height, weight, or capacity of something. Doing words, called verbs, like measure, have different forms.* Model the different forms of the word measure. Say, *I will measure the book.* Measure a book using connecting cube units. While measuring say, *I am measuring the book.* After you are finished measuring, say, *I measured the book.* Distribute connecting cubes to pairs. Have them measure a book and verbally use the different forms of the verb *measure*.	**Developing Oral Language** Use paper clips to measure the length of a desk. Say, *I will measure this desk. I will use paper clips to measure.* Emphasize the word measure. Count the paper clips. Say, *The desk measured _____ paper clips.* Distribute paper clips or pennies to pairs. Have them measure a desk or a book. Then have them report to the whole group using the following sentence frames: I measured my _____ (item). **I used paper clips/pennies to measure it. The (item) measured (number of units) paper clips/pennies.**

Teacher Notes:

NAME _____ DATE _____

Lesson 3 Vocabulary Definition Map
Nonstandard Units of Length

Use the definition map to write about the math word.

My Math Word:

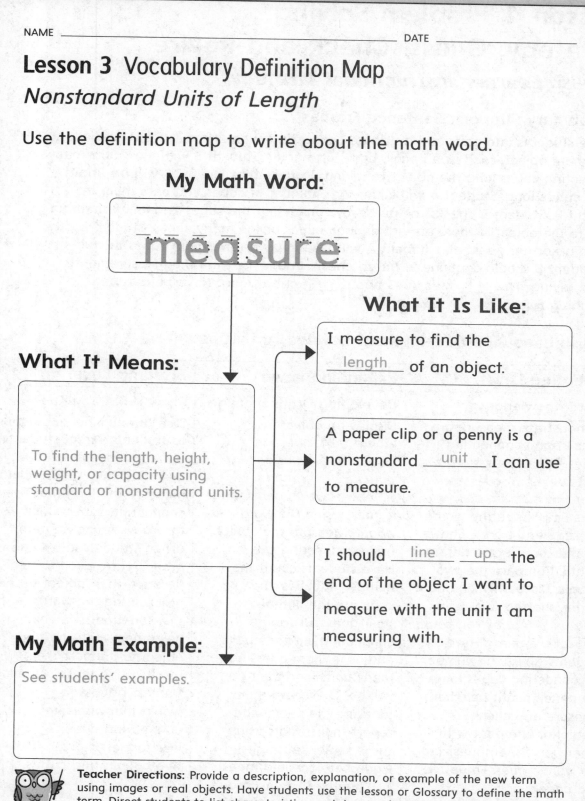

measure

What It Is Like:

I measure to find the ____length____ of an object.

What It Means:

To find the length, height, weight, or capacity using standard or nonstandard units.

A paper clip or a penny is a nonstandard ____unit____ I can use to measure.

I should ____line____ ____up____ the end of the object I want to measure with the unit I am measuring with.

My Math Example:

See students' examples.

Teacher Directions: Provide a description, explanation, or example of the new term using images or real objects. Have students use the lesson or Glossary to define the math term. Direct students to list characteristics, and draw a picture representing their math term. Then encourage students to describe their picture to a peer.

Grade 1 • **Chapter 8** *Measurement and Time* **83**

Lesson 4 Problem Solving
Strategy: Guess, Check and Revise
English Learner Instructional Strategy

Vocabulary Support: Sentence Frames

Divide students into pairs and distribute connecting cubes. Have student A select a classroom object, such as a pencil, book, or marker. Student B will guess how many connecting cubes long the object is, saying: **I guess that the _____ will be about _____ cubes long.** Student A will count that number of cubes and pass them to student B. Student B checks the guess, by connecting the cubes and lining them up next to the object. If there are not enough cubes or too many cubes, student B can revise his or her guess and student A will take away or give more cubes to student B. Student B should continue to guess, check, and revise until the measurement is found saying: **The _____ measures about _____ cubes long. Have pairs repeat** switching roles.

English Language Development Leveled Activities

Emerging Level	Expanding Level	Bridging Level
Word Knowledge	**Developing Oral Language**	**Show What You Know**
Demonstrate measuring a pencil using connecting cubes. Say, *I measured the pencil. The pencil is about seven connecting cubes long.* Emphasize the words *measure* and *about*. Show another pencil of a different length. Compare the two pencils. Discuss which one is longer/shorter. Ask, *How long do you think this pencil is? Make a guess.* Have students guess the number of connecting cubes long the pencil might be. Then measure and check predictions. Repeat with other pencils, emphasizing the words: *guess, measure* and *check*.	Display an object. Ask, *About how many connecting cubes would equal the length of this object?* Have students make a guess using the word *about*. Measure the object using connecting cubes. Have students count the number of cubes and compare the number to their guess. Distribute connecting cubes to pairs and have them guess how many units long their pencil will be. Then have them measure the pencil and report their results using these sentence frames: **I guessed it would measure _____ units. It measured _____ units.**	Distribute connecting cubes to students. Direct students to find an object at their desk that they guess would measure the length of five connecting cubes. Once they have found an item, divide students into small groups. Have each student describe their object to the group using the words, *measure* and *unit*. Then have them measure the object and report their results using these sentence frames: **I guessed _____ would measure five connecting cubes. _____ was longer/shorter than five connecting cubes.**

Teacher Notes:

NAME _____ DATE _____

Lesson 4 Problem Solving

STRATEGY: Guess, Check, and Revise

<u>Underline</u> what you know. (Circle) what you need to find. Guess and measure.

1. **About** how <u>many pennies long is the **object**?</u>

   penny

   Guess: about _____ pennies.

   Measure: about _____ pennies.

The **object** is **about** __3__ pennies long.

Teacher Directions: Provide a description, explanation, or example of the boldface terms and nouns using images or real objects. Read each sentence and have students echo read. Encourage students to guess how many pennies long the object is. Then allow them to use actual pennies to measure the object. Finally, have them write their answer in the restated question. Have students read the answer sentence aloud.

Lesson 5 Time to the Hour: Analog

English Learner Instructional Strategy

Vocabulary Support: Make Connections

Point to your face and say, *face*. Have students do the same. Say, *A face can show emotions depending on how it moves.* Display different emotions on your face (happy, sad, mad) and have students identify the emotion. Hold up your hands and say, *hands*. Have students do the same. Say, *Hands can point to things. Point to different classroom objects and have students identify the objects using the sentence frame:* **Your hand is pointing to the _____.**

Point to an analog clock's face and say, *face. The clock's face shows time.* Set 2:00 on the clock. Point to the clock's hands and say, *hands. Clock hands point to numbers.* Set times to the hour and have students identify where the hands are pointing using the sentence frame: **The hands are pointing to twelve and _____.**

English Language Development Leveled Activities

Emerging Level	Expanding Level	Bridging Level
Word Knowledge	**Listen and Identify**	**Act It Out**
Point to your face and have students do the same. Say, *face*. Have students repeat chorally. Point to your hands and have students do the same. Say, *hands*. Have students repeat chorally. Show students an analog clock. Introduce the face and hands of an analog clock to students. Discuss the difference between the hour hand and minute hand. Model how to set the hands to show 2:00 and say, *The clock shows two o'clock.* Encourage students to chorally repeat. Repeat with other times to the hour.	Use an analog clock to identify *analog clock, face, hour hand* and *minute hand*. Count the hour increments. Position the hands at 2:00. Say, *The clock shows two o'clock.* Discuss the difference between the hour and minute hands. Distribute analog clocks to students and ask them to set their clocks to any time to the hour. Have students describe their clock using the sentence frame: The minute hand is pointing at the twelve. **The hour hand is pointing at the _____. My clock shows _____ o'clock.**	Use construction paper to write the numbers 1–12 and place the numbers in a large circle on the floor to create an analog clock. Cut a large black circle and place it in the center of the clock. Make one red cone hat (hour hand) and one blue cone hat (minute hand) and have student volunteers wear them. Confidentially tell the pair to position themselves, on the floor, to show the time 3:00. Ask, *What time does our human analog clock show?* **3:00** Repeat with other times to the hour with new volunteers.

Teacher Notes:

NAME _____ DATE _____

Lesson 5 Vocabulary Sentence Frames
Time to the Hour: Analog

The math words in the word bank are for the sentences below. Write the words that fit in each sentence on the blank lines.

Word Bank		
o'clock	minute hand	hour hand

1. The long hand on the clock is the
   <u>minute</u>   <u>hand</u>.

2. It is two <u>o'clock</u>.

3. The hand that is pointing to 2 is
   the <u>hour</u>   <u>hand</u>.

Teacher Directions: Provide a description, explanation, or example of the each term using images or real objects. Read each sentence frame and have students echo read. Direct students to write the correct terms in each blank. Then encourage students to read each sentence to a peer.

Grade 1 • **Chapter 8** *Measurement and Time* **85**

Lesson 6 Time to the Hour: Digital

English Learner Instructional Strategy

Collaborative Support: Act It Out

Write the numbers 1–12 on pieces of paper. Distribute the papers to 12 students. Say, *Each of you is holding a different hour.* Write 00 on a piece of paper. Distribute the 00 paper to a student. Say, *You are the minutes.* Write a large colon on a piece of paper. Distribute the colon paper to a student. Say, *You are the clock's colon that separates the hours from the minutes.*

Explain that you are going to say aloud a time to the hour and three students will come to the front of the room and create that time in digital form (the hour, the colon, and the minutes). Repeat the activity. Say aloud various times to the hour and have different students hold the hour, the colon, and 00 minutes papers to mimic a digital clock. Continue until all students have had a chance to participate.

English Language Development Leveled Activities

Emerging Level	Expanding Level	Bridging Level
Listen and Identify	**Show What You Know**	**Show What You Know**
Display and review the parts of an analog clock: face, minute and hour hands. Review times to the hour (2:00, 5:00, and 8:00). Display a digital clock. Say, *This is a digital clock. The colon separates the hour and the minutes.* Discuss how the hour appears to the left of the colon, and the minutes appear to the right. Set both clocks to the same time to the hour. Have students identify time to the hour using the sentence frame: **The time is _____ o'clock.**	Display an analog clock. Ask students, *What kind of clock is this?* **analog clock** Set the clock to 2:00 and have students read the time. Display a digital clock. Say, *This is a digital clock.* Write *2:00* on the board. Explain the position of the hour (2) to the left of the colon, and the minutes (00) to the right of the colon. Provide pairs an analog clock and a write-on/ wipe-off board. One student displays a time to the hour and the other writes the time in digital form. Have students switch roles and repeat.	Display a digital clock. Ask a volunteer to summarize how it shows time differently from an analog clock. Provide pairs of students with a manipulative analog clock and write-on/ wipe-off board. Have one student show a time to the hour and the other student identify and write the time in digital form. Have students switch roles and repeat. Circulate around the pairs and have students report back the time to the hour using the following sentence frame: **This is _____ o'clock on the _____ (analog/digital) clock.**

Teacher Notes:

NAME _____ DATE _____

Lesson 6 Four-Square Vocabulary
Time to the Hour: Digital

Trace the words. Write the definition for *digital clock*. Write what the word means, draw a picture, and write your own sentence using the word.

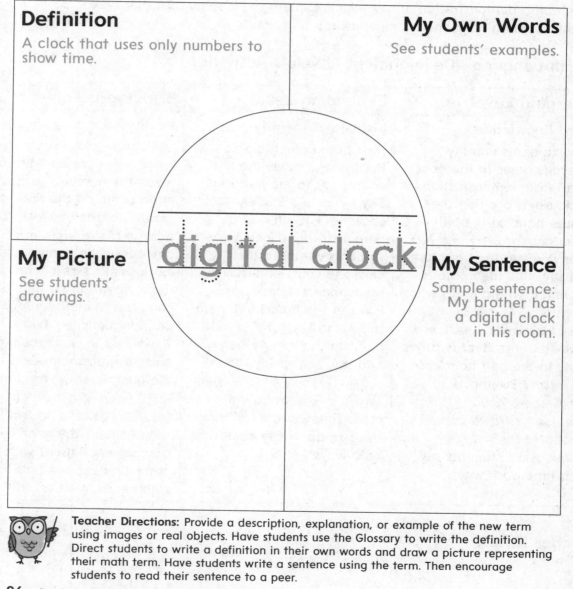

Definition
A clock that uses only numbers to show time.

My Own Words
See students' examples.

digital clock

My Picture
See students' drawings.

My Sentence
Sample sentence: My brother has a digital clock in his room.

Teacher Directions: Provide a description, explanation, or example of the new term using images or real objects. Have students use the Glossary to write the definition. Direct students to write a definition in their own words and draw a picture representing their math term. Have students write a sentence using the term. Then encourage students to read their sentence to a peer.

Lesson 7 Time to the Half Hour: Analog
English Learner Instructional Strategy

Vocabulary Support: Cognates

Write the words *hour/minute* and their Spanish cognates, *hora/minute* on a cognate chart. Ask, *What things can take an hour in length?* Accept and discuss all answers. Give examples of activities that last an hour, such as: music or gym class, piano lesson, TV show, recess, lunch, nap, car ride. Write the words *half hour*. Ask, *What are some things that are a half hour or thirty minutes in length?* Accept and discuss all answers. Give examples of activities that last a half hour, such as: coloring a picture, game, bike ride, breakfast, bath, baking cookies. Demonstrate an hour and half hour showing the minutes in each on a demonstration analog clock. Explain to students that there are 60 minutes in an hour and 30 minutes in a half hour.

English Language Development Leveled Activities

Emerging Level	Expanding Level	Bridging Level
Word Knowledge	**Listen and Identify**	**Act It Out**
Have students identify different times to the hour on a large demonstration clock. Next, position the minute hand to the half hour. Say, *This shows a half hour.* Point to the hour hand and say, *two* then point to the minute hand and say, *thirty.* Say, *Two-thirty. We can also say **half past** two.* Have students identify other times to the half hour such as: 5:30, 7:30 and 11:30. Say, *Eleven-**thirty** is the same as **half past** eleven. Eleven-thirty. Half past eleven.* Have students say each time both ways.	Distribute analog clocks. Discuss and model how to show time to the half hour. Say, *Show me 3:30.* Model showing 3:30 on a large demonstration clock. Have students compare their clocks to yours and make adjustments if necessary. Point to the hour hand near the 3 and say, *three*, and point to the minute hand and say, *thirty.* Say, *Three-thirty. We can also say half past three.* Repeat with other times to the half hour. Have students say each time both ways.	Use construction paper to write the numbers 1–12 and place the numbers in a large circle on the floor to create an analog clock. Cut a large black circle and place it in the center of the clock. Make a red cone hat (hour hand) and a blue cone hat (minute hand). Confidentially ask two volunteers to wear them and position themselves, on the floor, to show the time 7:30. Ask, *What time does our human analog clock show?* **seven-thirty or half past seven.** Repeat with other times to the half-hour with new volunteers.

Teacher Notes:

NAME _____ DATE _____

Lesson 7 Concept Web
Time to the Half Hour: Analog

Trace the term. Draw a line from the term to each clock that shows a half hour.

Teacher Directions: Provide a description, explanation, or example of the new term using images or real objects. Direct students to trace the term in the box. Then have students draw a line from the term to each picture that shows time to the half hour. Model and have students practice sentences to describe the time, such as: **It is eight-thirty**. or **It is half past eight**.

Grade 1 • **Chapter 8** *Measurement and Time* **87**

Lesson 8 Time to the Half Hour: Digital

English Learner Instructional Strategy

Collaborative Support: Act It Out

Write the numbers 1–12 on pieces of paper. Distribute the papers to 12 students. Say, *You twelve students are different hours.* Write 30 on a piece of paper. Distribute the 30 paper to a student. Say, *You are the minutes.* Write a large colon on a piece of paper. Distribute the colon paper to a student. Say, *You are the clock's colon that separates the hours from the minutes.*

Explain that you are going to say aloud a time to the half hour and three students will come to the front of the room and create that time in digital form (the hour, the colon, and the minutes). Repeat the activity. Say aloud various times to the half hour and have different students hold the hour, the colon and 30 minutes papers to mimic a digital clock. Continue until all students have had a chance to participate.

English Language Development Leveled Activities

Emerging Level	Expanding Level	Bridging Level
Listen and Identify	**Building Oral Language**	**Show What You Know**
Display an analog and a digital clock. Point to each one and ask, *Is this an analog clock? Is this a digital clock?* Have students respond with **yes/no** or thumbs-up/thumbs-down as appropriate. Set both clocks to 2:30. Say, *The time is half past two or two-thirty.* Have students repeat chorally. Continue setting both clocks to the same time on the half hour. Have students identify the time using this sentence frame. **The time is half past _____ or _____ thirty.**	Show a digital clock and ask, *What kind of clock is this?* **digital clock** Have students explain how they know. Repeat with an analog clock. Set both clocks to the same time to the half hour. Have students read the time on each clock. Next, set the digital clock to a time to the half hour and have students tell you how to set the hands on the analog clock to show the same time. Have students use the terms: *analog clock, digital clock, half past, thirty, o'clock, minute hand, hour hand.*	Show 3:30 on an analog clock. Demonstrate how a digital clock shows the same time. Remind students that the hour is before the colon and the minutes are after the colon. Distribute analog clocks and write-on/wipe-off boards. Say a time to the half hour and have each pair write the time in digital form. Then pairs represent the time on the analog clock. Have volunteers describe the time using this sentence frame: **This is _____ thirty or half past _____ on the (analog/digital) clock.**

Teacher Notes:

NAME _____ DATE _____

Lesson 8 Vocabulary Word Identification
Time to the Half Hour: Digital

Trace the words. Label the pictures with a term
from the word bank.

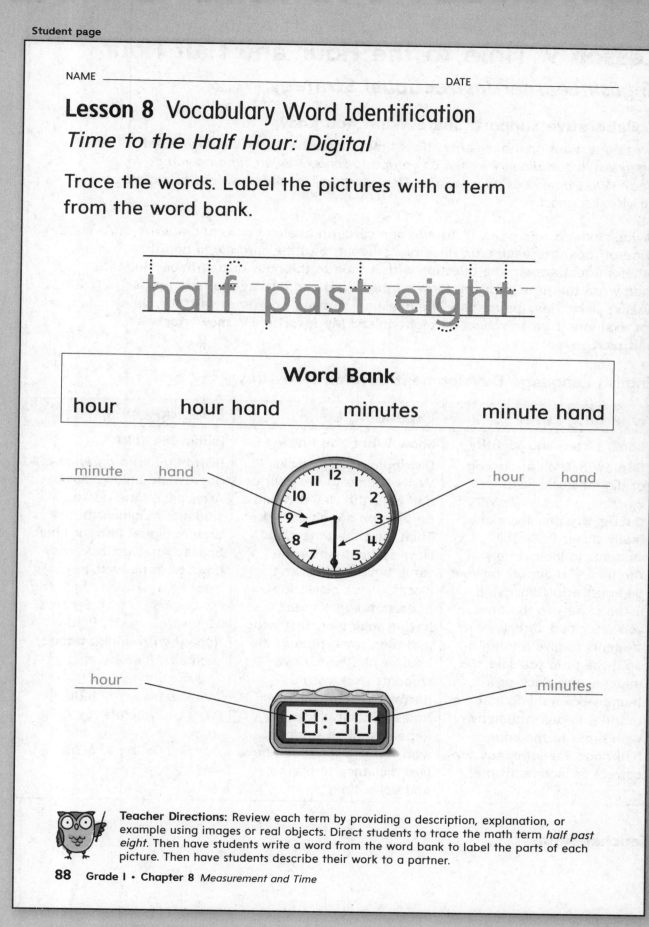

half past eight

Word Bank			
hour	hour hand	minutes	minute hand

minute ___ hand ___   hour ___ hand ___

hour ___   minutes ___

8:30

Teacher Directions: Review each term by providing a description, explanation, or
example using images or real objects. Direct students to trace the math term *half past
eight*. Then have students write a word from the word bank to label the parts of each
picture. Then have students describe their work to a partner.

Lesson 9 Time to the Hour and Half Hour

English Learner Instructional Strategy

Collaborative Support: Share What You Know

Write questions on index cards about time, such as: *What time do you eat breakfast/lunch/dinner? When do you go to school? What time do you go to bed? What time does your favorite TV show start?* Provide drawings or pictures for visual support.

Divide students into pairs. Distribute one card, an analog clock, and a write-on/wipe-off board to each pair. Have one student read the card aloud and the other student answer the question with a time to the hour or half-hour. Students then write the time in digital form on the board and display the time on the analog clock. Have pairs share the question, answer, and times with the class. For example, **I go to school at eight-thirty. My favorite TV show starts at half past seven.**

English Language Development Leveled Activities

Emerging Level	Expanding Level	Bridging Level
Look, Listen and Identify Display 8:30 on an analog or digital clock. Say, *My clock shows nine-thirty.* Discuss whether the clock really shows 9:30. Tell students to **look** closely at the time you display on your demonstration clock and **listen** closely to the time you say aloud. Direct students to give a thumbs-up if the time you said matches the clock or a thumbs-down if you made a mistake. Repeat the activity, with times to the hour or half hour. Randomly say the correct or incorrect time.	**Show What You Know** Distribute analog clocks. Write a time to the hour or half hour the way it would be seen on a digital clock. Then read the time aloud. Have students show the same time on their analog clocks. Check clocks for understanding. Repeat, having volunteers first write and then say a time to the hour or half hour. Have students work with a partner to compare the times on their clocks. Repeat until all students, who want to volunteer, have had a chance to present and say a time.	**Riddle Me This!** Distribute write-on/wipe-off boards and analog clocks. After each time riddle, students should write the time in digital form on their boards. Analog clocks may be used, if needed. Say, *My hour hand is between 9 and 10. My minute hand is on 6. What time am I?* **9:30** Repeat with similar riddles, such as: *The time on my clock is one hour after 7:00. What time am I?* **8:00** *I have a 6 before my colon and a 30 after my colon. What time am I?* **6:30**

Teacher Notes:

NAME _____ DATE _____

Lesson 9 Note Taking
Time to the Hour and Half Hour

Read the question. Write words you need help with. Use your lesson to write your Cornell notes. Write or draw math examples to explain your thinking.

Building on the Essential Question	Notes:
How can I tell time to the hour and half hour?	I know that one __hour__ is 60 minutes.
	The __hour__ __hand__ points to the hour.
	If the minute hand ends at 12, the clock shows an hour.
Words I need help with: See students' words.	I know that a __half__ __hour__ is 30 minutes.
	If the minute hand ends at 6, it shows 30 __minutes__. This is the half hour.
	A digital clock shows the __hours__ on the left and the __minutes__ on the right.

My Math Examples:
See students' examples.

Teacher Directions: Read the Building on the Essential Question and have students list words/phrases they need assistance with. Provide descriptions, explanations, or examples of the terms using images or real objects. Read each sentence frame and have students write the appropriate terms. Have students read their notes aloud. Direct students to draw a picture representing the question. Then encourage students to describe their picture to a peer.

Grade 1 • **Chapter 8** Measurement and Time **89**

Chapter 9 Two-Dimensional Shapes and Equal Shares

What's the Math in This Chapter?

Mathematical Practice 7: Look for and make use of structure

Present an image of a window made up of geometric panes. Ask, *What shapes do you see?* Allow students time to respond and discuss how they identified the shapes. Define *structure* as "a pattern of shapes or qualities organized to make a whole" using images of real-world 2-D shapes as visual support (e.g., window panes, flags, signs, targets). Model identifying two-dimensional shapes, structure, and equal shares within these examples.

Ask, *Have you ever noticed a picture made of shapes?* Solicit **Yes** from students, then have groups identify shapes within classroom charts or images.

Ask, *How can structure in math help us?* Have students discuss with a peer, then as a group. The goal is to recognize that understanding structure can help them identify shapes.

Display a chart with Mathematical Practice 7. Restate Mathematical Practice 7 and have students assist in rewriting it as an "I can" statement, for example: **I can see structure in shapes. I understand how shapes are put together as parts and wholes.** Have students draw two-dimensional shapes, and post the chart in the classroom.

Inquiry of the Essential Question:

How can I recognize two-dimensional shapes and equal shares?

Inquiry Activity Target: **Students come to a conclusion that structure can be used to understand: shape attributes, how shapes are put together, and how shapes can be partitioned into equal parts.**

As an introduction to the chapter, present the Essential Question to students. The inquiry graphic organizer will offer opportunities for students to observe, make inferences, and apply prior knowledge of two-dimensional shapes representing the Essential Question. As they investigate, encourage students to draw, write, and collaborate with peers to demonstrate their observations and thinking. Then have students present additional questions they may have to a peer to extend discussions.

Regroup students and restate Mathematical Practice 7 and the Essential Question. Pose questions to reflect on what has been learned to guide students in making connections between the Mathematical Practice and the Essential Question.

NAME _____ DATE _____

Chapter 9 Two-Dimensional Shapes and Equal Shares
Inquiry of the Essential Question:

How can I recognize two-dimensional shapes and equal shares?

3 Sides
3 Vertices

4 Sides
4 Vertices

I see ...

I think ...

I know ...

Rectangle

Rectangle

I see ...

I think ...

I know ...

I see ...

I think ...

I know ...

Questions I have...

_____

_____

_____

Teacher Directions: Read the Essential Question for students. Have students echo read. Direct students to describe their observations, inferences, and prior knowledge of each math example. Encourage students to write or draw additional questions they may have. Then have students share their ideas/questions with a peer.

Lesson 1 Squares and Rectangles
English Learner Instructional Strategy

Vocabulary Support: Modeled Talk

Create many different sizes and colors of squares and rectangles from pieces of construction paper. Draw examples of a rectangle and a square. Ask, *How are these shapes the same?* **Both have four sides and four vertices.** Label one side and one vertex on each shape. *How are they different?* **One has the same length sides.** Point to the square and say, *square*. Label with the word *square*. Point to the rectangle and say, *rectangle*. Label with the word *rectangle* and the Spanish cognate, *rectángulo*. Say, *A square is a special kind of rectangle that has four equal sides*. As you display each paper square or rectangle example, ask, *Is this a rectangle or a square?* Have students say, **square** or **rectangle**.

English Language Development Leveled Activities

Emerging Level	Expanding Level	Bridging Level
Word Knowledge	**Making Connections**	**Listen and Draw**
Write the word *rectangle*. Use clay and toothpicks to form a square. Have students repeat the following sentences chorally. Say, *This is a square.* Point to a ball of clay connecting two sides of the square and say, *This is a vertex.* Point to all four and say, *These are vertices.* Point to a toothpick and say, *This is a side.* Have students count the sides. Say, *A square has four equal sides.* Repeat to make a rectangle. Break toothpicks in half to represent the shorter sides.	Draw a square and a rectangle. Have students count the sides and then point to the vertices of each shape. Say, *Both shapes have four **sides** and four **vertices**.* Measure using a standard or non-standard form of measurement to show that a square has four equal sides in length and the opposite sides of a rectangle are equal in length. Ask, *How are these shapes different?* Have students use this sentence frame: **The ____ has ____.** Display different sized square and rectangle shapes. Have students describe each shape, using the sentence frame: **This is a ____ because it has ____.**	Draw a square and a rectangle on the board. Explain that these are two-dimensional shapes with length and width. Distribute write-on/ wipe-off boards. At your prompt, have students quickly draw a shape. For example, say, *Draw a rectangle!* or *Draw a Square!* Students then draw the shape and display their board. Check boards for understanding. Repeat the activity having students draw different configurations of squares and rectangles such as: *Draw 3 squares! Draw a square inside of a rectangle! Draw a little rectangle and a big rectangle!*

Teacher Notes:

NAME _____ DATE _____

Lesson I Word Web
Squares and Rectangles

Label each item with a word from the word bank.

Word Bank

rectangle	side	square	vertex

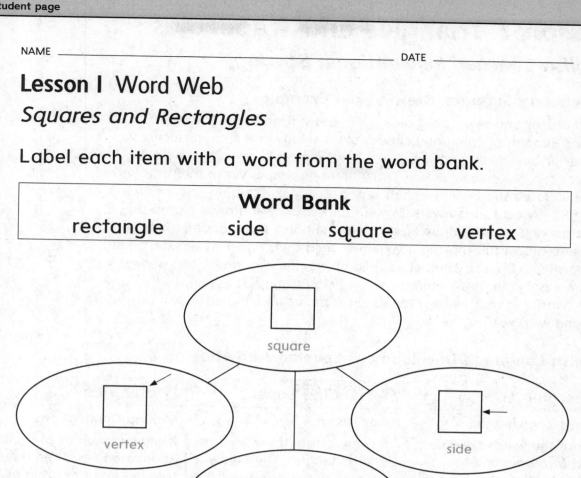

square

vertex

side

two-dimensional shapes

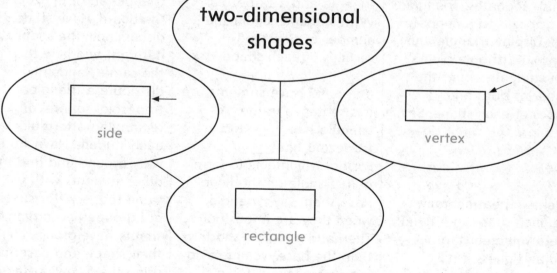

side

vertex

rectangle

Teacher Directions: Provide a description, explanation, or example of the new terms using images or real objects. Teach students that the plural of vertex is vertices. Direct students to look at each picture and then label it with the correct word from the word bank. Model and have students practice sentences that describe each shape, such as: **A square has four sides and four vertices.**

Grade I • **Chapter 9** *Two-Dimensional Shapes and Equal Shares* **91**

Lesson 2 Triangles and Trapezoids
English Learner Instructional Strategy

Vocabulary Support: Draw Visual Examples

Open a door and say, *The door is open*. Close it and say, *The door is closed*. Repeat, opening and closing other things in the room, such as a drawer or backpack, and say, *The ____ is open. The ____ is closed*. Say, *Shapes can be open or closed*. Draw an open shape. Write then say, *open*. Draw a closed shape. Write then say, *closed*. Ask, *How are these shapes different?* Accept all answers. Explain that when you draw a closed shape, you start and end the drawing at the same point. Draw another closed shape to model this, having a volunteer hold their finger at the start point of the shape. Discuss whether you did or did not stop where the student's finger is pointing. Draw more open and closed shapes, asking, *Is this a(n) open/closed shape?* Have students show thumbs-up/thumbs-down or respond with **yes/no**.

English Language Development Leveled Activities

Emerging Level	Expanding Level	Bridging Level
Word Knowledge	**Act It Out**	**Making Connections**
Write the words *triangle* and *trapezoid* and the Spanish cognates, *triangulo* and *trapezio* on a cognate chart. Display triangle and trapezoid attribute blocks. With students, count the number of sides and vertices of each shape. Say, *A triangle has three sides and three vertices. A trapezoid has four sides and four vertices*. Have students repeat chorally. Hide each shape in a hand. Have a volunteer tap one hand and guess if it contains a triangle or trapezoid. Reveal and say, *Yes, a triangle/trapezoid*, or *No, a trapezoid/ triangle*. Repeat with different volunteers.	Review attributes of squares and rectangles. Then draw a triangle and a trapezoid. With students, count the number of sides and vertices of each shape. Ask, *How are the shapes different?* Have students use this sentence frame: **A triangle has ____ and a trapezoid has ____.** Position volunteers to form of a triangle on the floor. Have them say, **triangle**, when they are in position. Then add a fourth (shorter than the base) volunteer to form a trapezoid. Have them say, **trapezoid**. Repeat with other students forming a rectangle and a square.	Review attributes of squares and rectangles. Then draw a triangle and trapezoid on the board. Have students discuss how the shapes are different and how they are the same. Randomly distribute a triangle, trapezoid, square, or rectangle pattern block to each student. Have students walk around and find the other students with the same shape. After students are grouped according to shape, have groups identify their shape and describe it using the following sentence frames: **This is a ____. It has ____ sides and ____ vertices.**

Teacher Notes:

NAME _____   DATE _____

Lesson 2 Vocabulary Word Identification
Triangles and Trapezoids

Match each word to a picture.

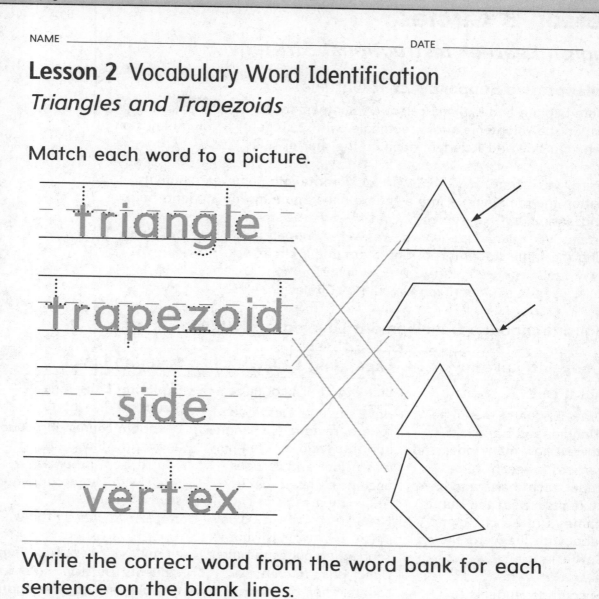

triangle

trapezoid

side

vertex

Write the correct word from the word bank for each sentence on the blank lines.

Word Bank			
sides	triangle	trapezoid	vertices

1. I have 4 __sides__ and 4 __vertices__. I am a __trapezoid__.

2. I have 3 __sides__ and 3 __vertices__. I am a __triangle__.

Teacher Directions: Review the terms using images or real objects. Have students say each term then draw a line to match the term to a picture showing its meaning. Direct students say each term in the Word Bank and discuss plural forms of side (sides) and vertex (vertices). Have students write the corresponding terms in the sentences. Encourage students to read the sentences to a peer.

92   Grade I • **Chapter 9** *Two-Dimensional Shapes and Equal Shares*

Lesson 3 Circles
English Learner Instructional Strategy

Collaborative Support: Act It Out

Before Explore and Explain, select 6 volunteers to act out the following. Position the volunteers into a rectangle, with 2 students holding hands on each long side and 1 student each on the shorter sides. Ask, *How many sides are in the shape we created?* **4** *How many vertices?* **4** *Is our shape closed?* **yes** *What shape did we create?* **rectangle** Remove 2 students. Position the remaining 4 into a square once again having students hold hands. Ask, *How many sides are in the shape we created?* **4** *How many vertices?* **4** *Is our shape closed?* **Yes** *What shape did we create?* **square** Using the same students, position them into a circle. Ask, *How many sides are in the shape we created?* **0** *How many vertices?* **0** *Is our shape closed?* **yes** *What shape did we create?* **Circle**

English Language Development Leveled Activities

Emerging Level	Expanding Level	Bridging Level
Act It Out	**Building Oral Language**	**Exploring Language Structure**
Draw a square, rectangle, triangle, and trapezoid. Review how many sides and vertices for each shape. Label each shape and have students repeat the shape names. Draw a circle and label with the word *circle*. Say, *This is a circle. No sides. No vertices.* Encourage students to repeat chorally. Direct students to stand, join hands and form a circle. Prompt commands, such as: *Walk in a circle. Sit in a circle. Hop in a circle.* Have students repeat each command chorally.	Have students identify a square, rectangle, triangle, and trapezoid and tell how many sides and vertices each has. Draw a circle. Ask, *How many sides?* **0** *How many vertices?* **0** Say, *This is a* **circle**. Distribute connecting cubes to pairs. Direct pairs to create a square. Then direct pairs to create a circle. Discuss why you cannot create a circle with the cubes connected. Distribute string to each pair. Direct pairs to create a circle with string. Explain that a circle has no sides and no vertices.	Distribute connecting cubes and string. Say, *Connect your cubes and create a circle.* Wait some time and say, *We **can't** use connected cubes to make a circle. We **cannot** use cubes to make a circle.* Next, have students use string to create a circle. Discuss sides and vertices. Have students use string to make other shapes. Discuss the difficulty in making shapes with vertices using string. Ask students to describe modeling shapes using this sentence frame: I _____ (can/can't/cannot) use _____ (string/connecting cubes) to make a _____ (shape name).

Teacher Notes:

NAME _____ DATE _____

Lesson 3 Vocabulary Definition Map
Circles

Use the definition map to write what the math word means and tell what the word is like. Write or draw a math example. Share your examples with a classmate.

My Math Word:

circle

What It Means:

A closed round shape.

What It Is Like:

A circle is ___closed___ and ___round___.

A circle does **not** have ___sides___.

A circle does **not** have ___vertices___.

My Math Example:

See students' examples

Teacher Directions: Provide a description, explanation, or example of the new term using images or real objects. Have students use the lesson or Glossary to define the math term. Direct students to list characteristics, and draw a picture representing their math term. Then encourage students to describe their picture to a peer.

Grade I · **Chapter 9** *Two-Dimensional Shapes and Equal Shares* **93**

Lesson 4 Compare Shapes
English Learner Instructional Strategy

Graphic Support: Graphic Organizer

Before the lesson, create a 5-column chart with the following headers: Shape, Name, Vertices, Sides, Equal Length Sides. Draw large rows on the chart. Create different sizes of the following shapes by cutting pieces of colored construction paper: square, rectangle, triangle, circle, trapezoid. Choose one of the paper shapes. Place the shape on the chart under the *Shape* column and complete the row as students answer these questions: *What is this shape's name? How many vertices? How many sides? Are any of the sides the same length?* Continue with the rest of the shapes. Place examples of the same shape in the same row. Using the completed chart, have students compare the shapes.

English Language Development Leveled Activities

Emerging Level	Expanding Level	Bridging Level
Word Knowledge	**Word Recognition**	**Show What You Know**
Draw a square, rectangle, triangle, trapezoid, and circle on the board. Review the sides and vertices. Identify and label each shape. Distribute a brown bag with one attribute block of any of the five shapes to each student. Divide student into pairs. Have Student A reach in his or her bag and feel the shape. Student A will guess the shape's name and pull it out to confirm. Student B then repeats the activity with his or her bag. Rotate bags to different pairs and repeat the activity.	Draw a square, rectangle, triangle, trapezoid, and circle on the board. Review the sides and vertices. Identify and label each shape. Distribute brown bags with attribute blocks of all five shapes to pairs of students. Students will take turns reaching into the bag to feel a shape. He or she will guess the shape without looking using the sentence frame: **This is a _____ because it has _____.** He or she will pull out the described shape and confirm with his or her partner.	Review the number of sides and vertices of a square, circle, triangle, trapezoid, and rectangle. Distribute brown bags containing attribute blocks of all five shapes to pairs of students. Have one student determine and describe the shape for the partner to find using this sentence frame: **Find a _____ with _____ sides and _____ vertices.** Have the other partner reach in the bag, feel around for the shape, and pull out the shape to confirm with their partner. Have pairs repeat the activity, switching roles each time.

Multicultural Teacher Tip

Individual praise and standing out from the crowd are frowned upon in some cultures. Students from these cultures may act embarrassed or uncomfortable when invited to the board to solve a problem or asked to share an answer with the class. They may be more comfortable during group classroom activities, and may prefer others in the group to share answers or demonstrate problem solving.

NAME _____ DATE _____

Lesson 4 Note Taking
Compare Shapes

Read the question. Write words you need help with. Use your lesson to write your Cornell notes. Write or draw math examples to explain your thinking.

Building on the Essential Question	**Notes:**
How can I compare two-dimensional shapes?	**Word Bank** length number round sides straight type I can check if the shapes are __round__ or if they have __straight__ sides. I can count the number of __sides__. I can check if the sides have the same __length__. I can count the __number__ of vertices. I can look for the same __type__ of shape.
Words I need help with: See students' words.	

My Math Examples:
See students' examples.

Teacher Directions: Read the Building on the Essential Question and have students list words/phrases they need assistance with. Provide descriptions, explanations, or examples of the terms using images or real objects. Read each sentence frame and have students write the appropriate terms. Have students read their notes aloud. Direct students to draw a picture representing the question. Then encourage students to describe their picture to a peer.

Lesson 5 Composite Shapes

English Learner Instructional Strategy

Sensory Support: Photographs/Realia

Gather images of bread, peanut butter, and jelly. Place each image on a paper equilateral triangle. Hold up each food triangle and ask, *What is this?* **bread; peanut butter; jelly** Say, *Some foods can be put together to make another food. When I put bread, peanut butter, and jelly together I can make a sandwich.* Present another example, using dried fruit, seeds, and nuts to make trail mix.

Flip the food triangles over to the blank side. Say, *Some shapes can be put together to make another shape. The new shape is called a* **composite shape.** Use the paper triangles to model how three triangles can make a trapezoid. Say, *When I put these three triangles together, I can make a* **composite shape.** *This* **composite shape** *is a* **trapezoid.**

English Language Development Leveled Activities

Emerging Level	Expanding Level	Bridging Level
Making Connections Display a square pattern block and say, *square.* Put two squares together to make a rectangle. Say, *Two squares can make a rectangle, a* **composite** *shape.* Have students repeat chorally. Display a triangle pattern block and say, *triangle.* Put three triangles together to make a trapezoid. Say, *Three triangles can make a trapezoid, a* **composite** *shape.* Have students repeat chorally. Display other composite shapes. Identify the shapes used to make the composite shape using the sentence frame: **____ and ____ make this composite shape.**	**Listen and Identify** Draw a square, rectangle, triangle, trapezoid, and circle. Label each shape and have students repeat the shape names. Distribute pattern blocks. Have students make a rectangle with two squares. Identify it as a *composite shape,* a shape made of other shapes. Then have students use this sentence frame to describe the shape: **This is a composite shape. It is made out of a ____ and a ____.** Have students make their own composite shapes with the blocks and describe it to a partner.	**Developing Oral Language** Distribute pattern blocks of squares and triangles. Direct students to put two squares together, making a rectangle. Say, *When you put two or more shapes together you make a new shape called a* **composite** *shape.* Have pairs play a composite shape game. Using a divider, such as a folder propped between them, have one student make a composite shape and describe it as the partner tries to recreate the composite shape. Pairs then remove the divider and compare their composite shapes.

Teacher Notes:

NAME _____ DATE _____

Lesson 5 Vocabulary Word Study
Composite Shapes

Circle the correct word to complete the sentence.

I. I can _____ shapes to make a composite shape.

(put together)   compare

Show what you know about the word:

composite

There are __9__ letters.

There are __4__ vowels.

There are __5__ consonants.

__4__ vowels + __5__ consonants = __9__ letters in all.

Draw a picture to show what the word means.

See students' examples.

Teacher Directions: Provide a description, explanation, or example of the new term using images or real objects. Read the sentence and have students circle the correct word. Direct students to count the letters, vowels and consonants in the math term, then complete the addition number sentence. Guide students to draw a picture representing their math term. Then encourage students to describe their picture to a peer.

Grade I • **Chapter 9** *Two-Dimensional Shapes and Equal Shares* **95**

Lesson 6 More Composite Shapes

English Learner Instructional Strategy

Collaborative Support: Pairs

Write the term *composite shape*. Ask, *What is a composite shape?* Direct students to the Glossary. Then have students chorally read aloud, **A shape created by putting two or more shapes together.**

Divide students into pairs. Give each pair two pattern block shapes, such as a square and a trapezoid. Ask pairs to combine the two shapes making a new shape. Have pairs trace their new shape onto a piece of paper. Once all pairs have traced their composite shape, place them on chart paper. Discuss how the same two shapes can create many different composite shapes. Repeat with other shape combinations such as, a trapezoid and a square or a triangle and a trapezoid.

English Language Development Leveled Activities

Emerging Level	Expanding Level	Bridging Level
Word Knowledge	**Listen and Identify**	**Developing Oral Language**
Create a composite shape using 3 or 4 pattern blocks. Say, *This is a* **composite shape**. Next, create a new composite shape using the same blocks. Say, *This is another composite shape*. Put away all but 1 of the blocks. Say, *This is not a composite shape*. Continue displaying examples and non-examples of composite shapes. Ask, *Is this a composite shape?* Have students answer saying, **yes/ no** or gesture a thumbs-up/ thumbs-down.	Distribute pattern blocks. Explain that as you build a composite shape, students will model the same composite shape using their own blocks. Describe and create a composite shape that has no gaps or overlaps. Use position words such as; *above, below, right, left,* and *on top of* when directing students to model the shape. Have students compare their shapes to your shape and describe how it is alike and different. Create another composite shape, describe it, and have students recreate and compare it to their own.	Distribute pattern blocks. Have each student in a pair create a composite shape using 1 square and 2 triangles. Have pairs compare how their shapes are the same and different using the shape names. For example, *My composite shape has two triangles and one square. I put the triangles on the top and bottom of the square.* Have each pair describe to the group how their composite shapes are the same and different. Repeat, having one partner choose the blocks to use for the composite shapes.

Teacher Notes:

NAME _____ DATE _____

Lesson 6 Concept Web
More Composite Shapes

Trace the word *shapes*. Circle the correct word.

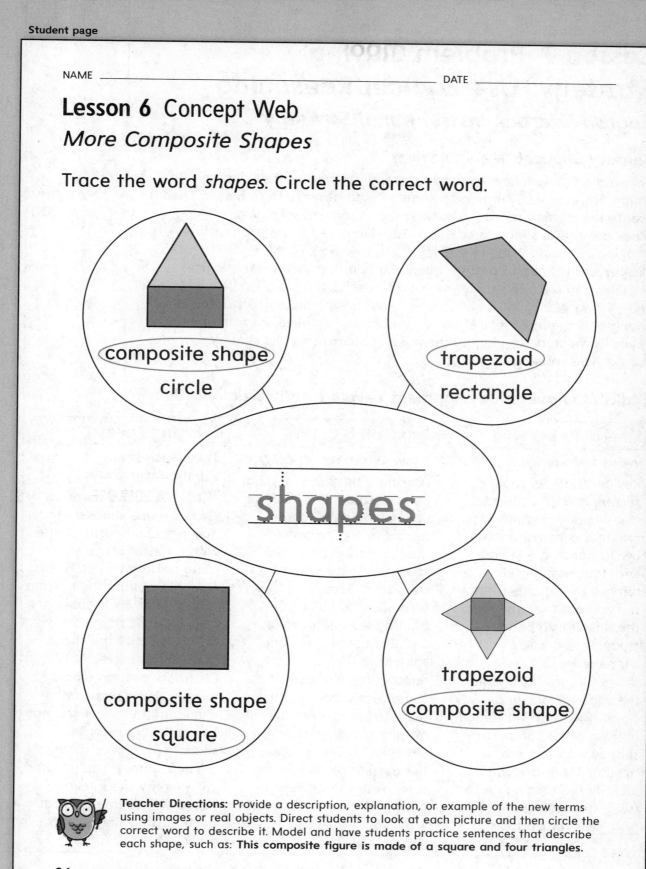

(composite shape)
circle

(trapezoid)
rectangle

shapes

composite shape
(square)

trapezoid
(composite shape)

Teacher Directions: Provide a description, explanation, or example of the new terms using images or real objects. Direct students to look at each picture and then circle the correct word to describe it. Model and have students practice sentences that describe each shape, such as: **This composite figure is made of a square and four triangles.**

96 Grade 1 • **Chapter 9** *Two-Dimensional Shapes and Equal Shares*

Lesson 7 Problem Solving
Strategy: Use Logical Reasoning

English Learner Instructional Strategy

Sensory Support: Manipulatives

Before the Problem Solving activity, prepare a missing block composite shape. For example, surround a triangle with other pattern blocks. Then create the missing part by removing the triangle. Display your missing block composite shape to students. Say, *I created a composite shape. A block is missing from my composite shape. Let's find the missing block.* Display each shape of pattern block. Explain that you will use logical reasoning to solve this problem. Point to each shape and ask, *Could this be the missing block? Why or why not?* Discuss why the correct shape (the triangle) is the most logical choice and place the block into the missing space. Have students identify the missing block using the sentence frame: **A _____ was missing.**

English Language Development Leveled Activities

Emerging Level	Expanding Level	Bridging Level
Word Knowledge	**Show What You Know**	**Think-Pair-Share**
Review *triangle, square, rectangle,* and *trapezoid*. Out of sight of students, create a composite shape by surrounding a square with four triangles and remove the square. Display the composite shape with the missing shape to students and ask, *What shape is missing from this composite shape?* Replace the square block and say, *A square was missing.* Repeat, making other composite shapes with one shape missing. Have students identify the missing piece using the sentence frame: **A _____ was missing.**	Distribute pattern blocks to pairs of students. Check for understanding of shape names by having pairs display a triangle, a square, a rectangle, and a trapezoid. Have pairs make a composite shape from 3 triangles, 2 trapezoids, and 2 squares. Have one student in the pair secretly remove one block from the composite shape and have the partner identify the missing piece. To check, have the student place the missing block into the composite shape. Repeat, having partners switch roles.	Ask, *What is a composite shape?* **A shape made of two or more shapes**. Model making a composite shape with a missing shape using pattern blocks. Have students identify the missing shape. After each guess, try to place that shape into the space until the shape guessed fits correctly. Distribute pattern blocks to pairs. Have one student challenge the other student by creating a composite shape with one or more missing shapes. The partner will identify the missing shape(s) by name and put each missing shape in its place.

Teacher Notes:

NAME _____ DATE _____

Lesson 7 Problem Solving
STRATEGY: Use Logical Reasoning

<u>Underline</u> what you know. (Circle) what you need to find. Use logical reasoning to solve.

I. <u>Rashad covered the pattern block with the **same three blocks.**</u>

(Circle which block he used.)

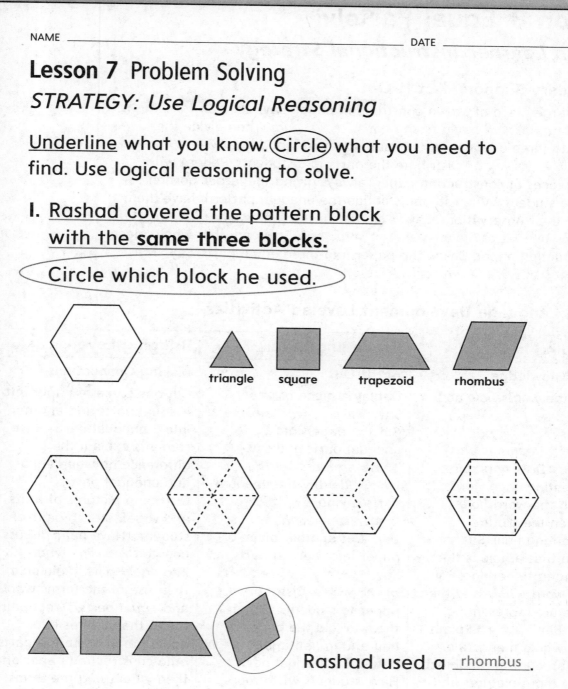

Rashad used a ___rhombus___ .

Teacher Directions: Provide a description, explanation, or example of the boldface terms and nouns using images or real objects. Read each sentence and have students echo read. Encourage students to use the patterns and then circle their answer. Model and teach sentence frames for describing students' work. For example, *I need two trapezoids to make the composite shape. That's not enough. I need six triangles to make the composite shape. That's too many.* Have partners practice describing the composite shapes.

Grade 1 · Chapter 9 *Two-Dimensional Shapes and Equal Shares* **97**

Lesson 8 Equal Parts

English Learner Instructional Strategy

Vocabulary Support: Act It Out

Show a large piece of brown construction paper. Say, *Let's pretend this is a chocolate bar. I'm going to share it with 3 friends.* Then divide the paper into three unequal parts (small, medium, and large). Say, *These are* **parts** *of the whole bar.* Distribute the parts to 3 students. Repeat with 2 more pieces of construction paper, always giving the larger portion to the same student. When students notice the unequal pattern, have them describe their observations. Say, *I did not divide the wholes into* **equal** *parts.* **Equal** *parts of a whole are the same size.* Show another whole piece of construction paper. Divide the paper into three equal parts. Say, *These are* **equal parts** *of the whole.*

English Language Development Leveled Activities

Emerging Level	Expanding Level	Bridging Level
Word Knowledge	**Act It Out**	**Making Connections**
Write the words *hole* and *whole*. Say, *These words sound the same, but have different meanings and spellings.* Display pictures representing each word. Point to the word *hole* and then the picture representing hole. Say, *hole.* Explain that a hole is the opening in something. Point to the word *whole* and then the picture representing whole. Say, *whole.* Explain that a whole means the entire amount of an object. Display more pictures and have students identify which word the picture represents.	Display a piece of paper. Say, *This is a* **whole** *piece.* Cut the paper into 3 unequal parts and say, *These are* **parts** *of the whole.* Identify the unequal parts saying, *This part is bigger/smaller than that part.* Cut another piece of paper into 3 equal parts. Say, *These are* **equal** *parts of the whole.* Distribute paper to students. Direct them to fold the paper in half and in half again. Count the 4 equal parts. Have students write *four equal parts* on the paper. Repeat with a circle, folding it into 2 equal parts.	Cut one piece of paper into 4 equal parts and another into 3 unequal parts. Have students explain the difference between equal and unequal parts. Distribute pictures of fruits and vegetables to pairs of students. Have pairs discuss how to divide the whole into equal parts. Reinforce their use of the terms *whole* and *equal parts.* Direct pairs to cut the pictures into equal parts, paste the parts onto construction paper and then label using the terms *whole* and *equal parts.*

Teacher Notes:

NAME _____ DATE _____

Lesson 8 Vocabulary Word Identification
Equal Parts

Trace then match each word to the picture.

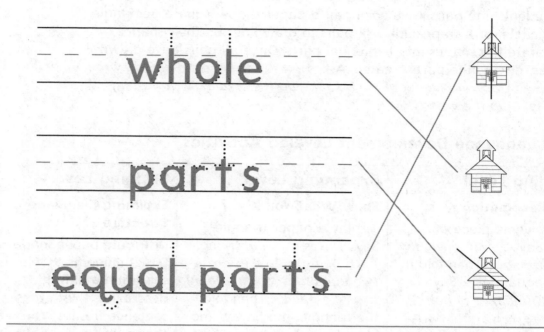

whole

parts

equal parts

Write the correct words from above for each sentence on the blank lines.

1. A whole can be separated into ___equal___ ___parts___.

2. Equal parts of the ___whole___ are the same size.

3. Sometimes, ___parts___ are not equal.

Teacher Directions: Review the new terms using images or real objects. Have students say each term and then draw a line to match each term to its meaning. Direct students to write the correct terms in the sentences. Encourage students to read the sentences to a peer.

98 Grade 1 · **Chapter 9** *Two-Dimensional Shapes and Equal Shares*

Lesson 9 Halves

English Learner Instructional Strategy

Collaborative Support: Pairs

Write the terms *halves* and *equal parts*. Say, *A whole that is separated into two equal parts is separated into halves.* Model dividing a whole into halves using an image of a whole sandwich or piece of round fruit. Say, *I will cut the _____ into two equal parts. I will cut the _____ into halves.* Divide students into pairs. Give each pair a paper shape (square, rectangle, triangle, circle, or a trapezoid). Ask pairs to draw lines on their shapes that separate the shapes into two equal parts. Once all pairs have drawn their lines, discuss the parts created. Ask, *How many lines did you draw on your shape?* **1** *How many **equal parts** make up the whole?* **2** *How many **halves** make up the whole?* **2**

English Language Development Leveled Activities

Emerging Level	Expanding Level	Bridging Level
Word Recognition Show a whole piece of paper. Say, *This is a **whole** piece of paper.* Then fold it in two equal parts. Say, *I folded the paper in **half**. There are two equal parts. The parts are called **halves**.* Have students repeat. Write *half* and *halves* on the board. Distribute a square, a rectangular and a circle shaped piece of paper to each student. Have students fold each piece in half. Have them write *halves* on each shape and use this sentence frame to describe their papers: **This paper is folded into halves.**	**Show What You Know** Display a paper triangle. Say, *This is a **whole** triangle. I am going to fold it into two equal parts. I am going to make **halves**.* Then fold the triangle into halves and identify each part as a half of the whole. Briefly discuss the singular and plural forms, *half* and *halves.* Distribute pieces of paper in different shapes, including circles, squares, rectangles, triangles, and trapezoids, to each student. Have students fold the shape they are given in half. Have students find other students who folded the same shape in half and have students match their shapes.	**Exploring Language Structure** Distribute paper shapes. Direct students to fold the shapes into halves, describing it with these sentence frames: **The whole shape was a _____. I folded it in half. Now I have two halves.** Have students label each side with the word *half* and write *halves* across the fold. Explain that the plural of *half* is *halves* and to spell it you change the *f* to *ve* and add *s*. On a T-Chart, list other irregular nouns that follow the same spelling changes when pluralized, such as: *elf, calf, knife, leaf, loaf, shelf,* and *wolf.*

Teacher Notes:

NAME _____ DATE _____

Lesson 9 Four-Square Vocabulary
Halves

Trace the word. Write the definition for *halves*.
Write what the word means, draw a picture, and
write your own sentence using the word.

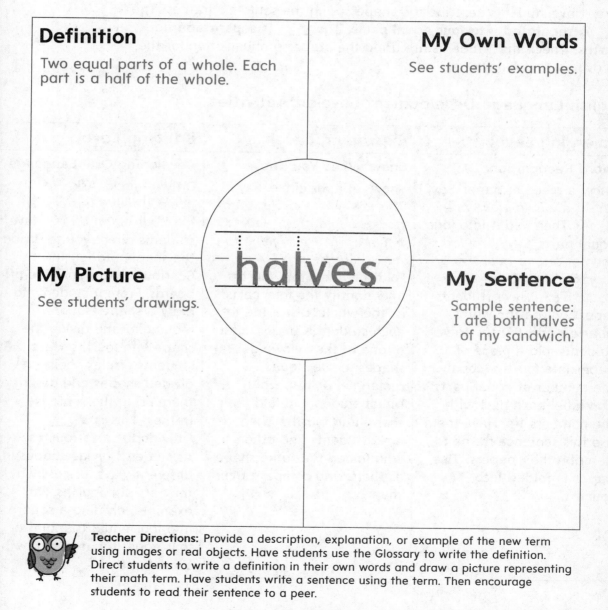

Definition

Two equal parts of a whole. Each
part is a half of the whole.

My Own Words

See students' examples.

halves

My Picture

See students' drawings.

My Sentence

Sample sentence:
I ate both halves
of my sandwich.

Teacher Directions: Provide a description, explanation, or example of the new term
using images or real objects. Have students use the Glossary to write the definition.
Direct students to write a definition in their own words and draw a picture representing
their math term. Have students write a sentence using the term. Then encourage
students to read their sentence to a peer.

Grade 1 • **Chapter 9** *Two-Dimensional Shapes and Equal Shares* **99**

Lesson 10 Quarters and Fourths

English Learner Instructional Strategy

Language Structure Support: Choral Responses

Draw a rectangle on the board. Draw a line dividing the rectangle into two equal parts. Say, *This rectangle is separated into **halves***. Then draw 2 more lines dividing the rectangle into four equal parts. Say, *The rectangle is now separated into **four** equal parts. The rectangle is separated into **fourths***. Write the words *four* and *fourths*. Say each word aloud as you point to it. Emphasize the "ths" sound in the word *fourths*. Draw other shapes and draw lines dividing the shapes into four equal parts. Have students describe the shapes using the sentence frames: **The ____ is separated into four equal parts. The ____ is separated into fourths.** Check that students are using the proper pronunciation for the words *four* and *fourths*.

English Language Development Leveled Activities

Emerging Level	Expanding Level	Bridging Level
Word Recognition Show a piece of paper. Say, *This is a whole piece of paper*. Then fold it into four equal parts. Say, *I folded the paper into **quarters** or **fourths**. There are four equal parts*. Have students repeat. Write the words *quarters* and *fourths*. Have students fold a piece of paper into fourths, count the number of equal parts, and label each part with the numbers 1–4. Have them use this sentence frame to describe their papers: **This paper is folded into fourths.**	**Show What You Know** Show a paper circle. Say, *This is a whole circle. I am going to fold it into four equal parts. I am going to make fourths or quarters*. Fold the circle into fourths and identify the four parts as the whole. Distribute to each student a paper cutout of one of the following shapes: circle, square, rectangle, or trapezoid. Direct students to fold their shape into fourths. Then have students find others who folded the same shape as theirs and compare their shapes.	**Developing Oral Language** Draw a circle. Ask, *How could I divide this circle into four equal parts?* Have students direct you to divide the shape into fourths. Distribute write-on/wipe-off boards. Prompt students to draw a shape (square, rectangle) and divide the shape into fourths. Have students display their divided shapes and describe them using the sentence frames: **This is a ____. I divided it into fourths.** As students divide shapes in different ways, discuss why they are still fourths. For example, dividing a square into four equal triangles or four equal squares can both show fourths.

Teacher Notes:

NAME _____ DATE _____

Lesson 10 Vocabulary Sentence Frames
Quarters and Fourths

The math words in the word bank are for the sentences below. Write the words that fit in each sentence on the blank lines.

Word Bank		
quarters	fourths	halves

1. ___Fourths___ are the four equal parts of a whole.

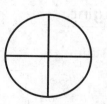

2. Fourths are also called ___quarters___.

3. ___Halves___ are two equal parts of a whole.

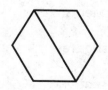

Teacher Directions: Provide a description, explanation, or example of the each term using images or real objects. Read each sentence frame and have students echo read. Direct students to write the correct terms in each blank. Then encourage students to read each sentence to a peer.

100   **Grade I • Chapter 9** *Two-Dimensional Shapes and Equal Shares*

Chapter 10 Three-Dimensional Shapes

What's the Math in This Chapter?

Mathematical Practice 7: Look for and make use of structure

Create and then present a composite object made of at least one cube, cylinder, and cone. Ask, *What shapes do you see?* Allow students time to respond. Remind them of the definition of structure. Say, *The structure of this object includes a cube, cylinder, and cone.* Then dismantle the object, identifying each three-dimensional shape.

Ask, *Have you ever noticed that objects are made of shapes?* Solicit **Yes** from students. Display a three-dimensional shape and ask, *Can you find an object in our classroom that has this structure?* Allow students time to find real-world objects that match, then repeat with another three-dimensional shape.

Ask, *How can structure in math help us?* Have students discuss with a peer, then as a group. The goal is to recognize that understanding structure can help them identify three-dimensional shapes.

Display the chart with Mathematical Practice 7 from Chapter 9. Restate Mathematical Practice 7 and have students assist in rewriting an additional "I can" statement, for example: **I can see the structure of shapes in real objects.** Have students find and cut out images of three-dimensional shapes and add them to the chart.

Inquiry of the Essential Question:

How can I identify three-dimensional shapes?

Inquiry Activity Target: **Students come to a conclusion that structure can be used to understand the shape of real-world objects.**

As an introduction to the chapter, present the Essential Question to students. The inquiry graphic organizer will offer opportunities for students to observe, make inferences, and apply prior knowledge of three-dimensional shapes representing the Essential Question. As they investigate, encourage students to draw, write, and collaborate with peers to demonstrate their observations and thinking. Then have students present additional questions they may have to a peer to extend discussions.

Regroup students and restate Mathematical Practice 7 and the Essential Question. Pose questions to reflect on what has been learned to guide students in making connections between the Mathematical Practice and the Essential Question.

NAME _____ DATE _____

Chapter 10 Three-Dimensional Shapes
Inquiry of the Essential Question:

How can I identify three-dimensional shapes?

Triangle

Rectangle

I see ...

I think ...

I know ...

Vertex

Face

Face

Vertex

I see ...

I think ...

I know ...

Cone

Cone

I see ...

I think ...

I know ...

Questions I have...

_____

_____

_____

_____

Teacher Directions: Read the Essential Question for students. Have students echo read. Direct students to describe their observations, inferences, and prior knowledge of each math example. Encourage students to write or draw additional questions they may have. Then have students share their ideas/questions with a peer.

Lesson 1 Cubes and Prisms
English Learner Instructional Strategy

Sensory Support: Modeled Talk

Show a cube (Spanish cognate *cubo*) and a rectangular prism (Spanish cognate *prismo retángulo*). Identify each by name and add the terms to the cognate chart with visual examples.

Show an open cube net. Ask, *Is this a cube?* Students should recognize that it is not yet a cube. Identify each *side, face,* and *vertex* (Spanish cognate *vértice*) as you fold the shape to make a cube. Count the number of faces and vertices. **6; 8** Repeat with a rectangular prism net.

English Language Development Leveled Activities

Emerging Level	Expanding Level	Bridging Level
Listen and Identify	**Make Connections**	**Vocabulary Game**
Show a cube and a rectangular prism. Identify each by name. Distribute one cube and one rectangular prism to each student. To the tune of "Mary Had a Little Lamb" sing: *Can you find a little cube, little cube, little cube? Can you find a little cube? Please, show it to me, now.* Have each student select the cube, hold it up, and say, **cube.** Repeat singing the song alternating between little cube and rectangular prism. Have students display the shape as you sing aloud.	Show a cube and rectangular prism. Identify each shape by name. Describe each one as being a three-dimensional shape. Compare the solids to a square and a rectangle. Show that the cube and prism have length, width, and height. Explain to students that they are going to go on a Shape Hunt. Have groups of students look in the classroom for examples of cubes and rectangular prisms. Have each group describe or draw the objects they identify and share with another group.	Write the following words on index cards: *rectangle, square, triangle, circle, rectangular prism,* and *cube.* Draw an example of each on additional cards. Create a set for each pair of students. Demonstrate how to play the "memory" game. Shuffle the cards and place facedown into rows. Players take turns flipping over two cards. If the cards create a match (show a shape and matching name), they are set aside. If the cards are not a match, they are flipped facedown. Play until all cards are matched. The player with the most matches wins.

Teacher Notes:

NAME _____ DATE _____

Lesson I Multiple Meaning Word
Cubes and Prisms

Trace the word. Say the math word. Draw a picture that shows the math word meaning in the first box. Then draw a picture that shows a non-math word meaning in the other box.

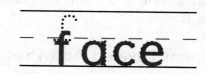

Math Meaning	Non-Math Meaning
Students' examples should represent: The flat part of a three-dimensional shape.	See students' examples.

Use the sentence frame below to help you describe your pictures.

This picture for the word ___face___ shows _____.

Teacher Directions: Provide math and non-math descriptions, explanations, or examples of the new term using images or real objects. Have students say then write the term. Then direct students to draw pictures showing a math and non-math meaning of the math term. Encourage students to describe their pictures to a peer using the sentence frame.

Lesson 2 Cones and Cylinders

English Learner Instructional Strategy

Vocabulary Support: Math Word Wall

Add the terms, *three-dimensional shape, cone, cylinder, cube, rectangular prism, face,* and *vertex* to the math word wall.

Frontload the following terms and solid shapes before the lesson. Write the word *face*. Say, *The flat part of a three-dimensional shape is a **face**. Some three-dimensional shapes have one face, and others have many more.* Display a cone and count its faces. **1** Display a cylinder and count its faces. **2** Write the word *vertex*. Say, *The **vertex** of a three-dimensional shape is a **point** where faces meet.* Display a cone and count the vertices. **1** Display a cylinder and count the vertices. **0**

English Language Development Leveled Activities

Emerging Level	Expanding Level	Bridging Level
Listen and Identify	**Making Connections**	**Act It Out**
Show a cone and a cylinder. Identify each and write the name. Point to the cone and say, *A cone has one face. A cone has one vertex.* Point to the cylinder and say, *A cylinder has two faces. A cylinder has no vertices.* Distribute geometric solids. Give attributes of a figure in the form of a riddle and have students guess what it is. For example, *I have six equal faces. What am I?* **cube** Have students respond using the sentence frame: **You are a _____ .**	Display a cone and a cylinder. Say, *A cone has one face. A cylinder has two faces.* Identify the faces. Next, say, *A cone has one vertex. A cylinder has zero vertices.* Identify the vertex and lack of vertex. Distribute a cylinder, a cone, a rectangular prism, and a cube to each student. Provide clues and have students choose the shape(s) that match your clues. For example, say, *These shapes have circular faces.* **cone and cylinder** Repeat with other descriptions.	Display a cylinder, cone, rectangular prism, and cube. Define them as three-dimensional shapes. Name and discuss the vertices and faces of each shape. Have groups play a guessing game. Demonstrate the game. Say, *I will think of a shape. Ask me questions to learn which shape I am thinking of.* Prompt students with, *I'm thinking of a three-dimensional shape.* Each student will ask a question such as, **Do you have one face?** or *Do you have zero vertices?* and then guess the shape.

Teacher Notes:

NAME _____ DATE _____

Lesson 2 Vocabulary Word Identification
Cones and Cylinders

Match.

cylinder

face

cone

vertex

Write the correct word from the word bank for each sentence on the blank lines.

Word Bank
cone   cylinder   face/faces   vertex/vertices

1. I have 2 __faces__ and 0 __vertices__. I am a __cylinder__.

2. I have 1 __face__ and 1 __vertex__. I am a __cone__.

Teacher Directions: Review the shapes using images or real objects. Have students say each shape name then draw a line to match the word to its corresponding image. Direct students say each term in the word bank, then write the corresponding terms in the sentences. Encourage students to read the sentences to a peer.

Grade 1 • **Chapter 10** *Three-Dimensional Shapes*   **103**

Lesson 3 Problem Solving Strategy: Look for a Pattern

English Learner Instructional Strategy

Sensory Support: Hands-On Activity

Before the lesson, model a pattern that is missing a shape using the online Geometric Solids, Virtual Manipulatives. Distribute geometric solids to each student, providing 2 of each shape. Create a pattern and then say the shape names aloud, such as: *Cone, cone, cylinder, rectangular prism, _____, cone, cylinder, rectangular prism.* Have students create the pattern with their solid blocks, then identify the missing shape. Prompt students to use the following sentence frames: **The missing shape is _____. The pattern is _____.**

English Language Development Leveled Activities

Emerging Level	Expanding Level	Bridging Level
Word Knowledge	**Think-Pair-Share**	**Show What You Know**
Review these three-dimensional shapes: cone, cylinder, cube, and rectangular prism. Display examples. Have students identify each shape. Use pictures of three-dimensional shapes to display a pattern. For example, 2 cylinders, 1 cone, 2 cylinders, 1 cone, 2 cylinders. Ask, *Which shape comes next?* **cone** Have a volunteer position a picture of a cone to complete the pattern. Repeat, using other pictures to create new patterns. Allow students to create patterns as well.	Glue 4 pictures of the following shapes on separate index cards to make a set of 16 cards: cube, cone, cylinder, rectangular prism. Create a set for each pair. Review the three-dimensional shapes: cone, cylinder, cube, and rectangular prism. Distribute a set of 16 cards to each pair of students. Have pairs take turns using the cards to create a pattern of shapes. Partners must identify the pattern by naming the shapes and extend the pattern by naming the shape that comes next.	Distribute 4 index cards with pictures of a cube, cone, cylinder, and rectangular prism to each student. Review the three-dimensional shapes on the cards. Have 5 volunteers stand at the front of the room with their cards. Have the volunteers display cards to show a pattern you describe. For example: *cube, cube, cone, cube, cube, _____.* Ask, *What shape comes next?* The seated students should display the correct shape (cone) and say, **A cone comes next in the pattern.** Repeat with different students and new patterns.

Teacher Notes:

NAME _____ DATE _____

Lesson 3 Problem Solving
STRATEGY: *Look for a Pattern*

<u>Underline</u> what you know. (Circle) what you need to find. Find a pattern to solve.

I. <u>Jenica made this pattern.</u>

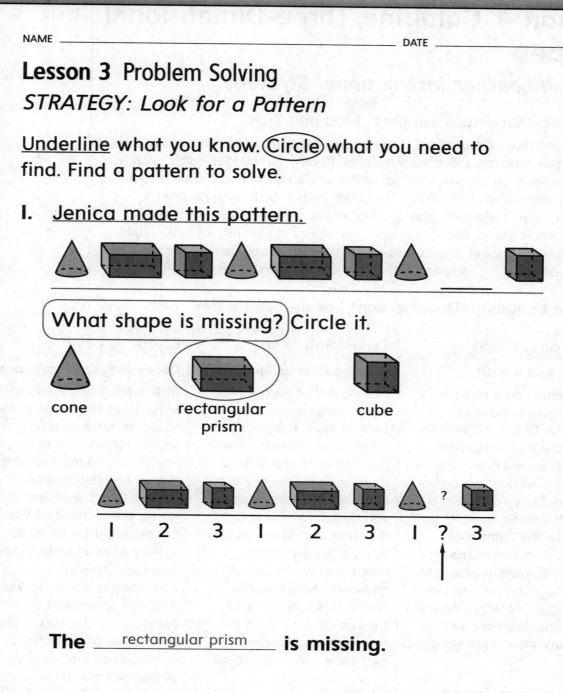

What shape is missing? Circle it.

cone   rectangular prism   cube

The _____rectangular prism_____ is missing.

Teacher Directions: Provide a description, explanation, or example of the bold face terms and nouns using images or real objects. Read each sentence and have students echo read. Encourage students to look for a pattern, discover the missing shape, circle the correct answer, and then write the shape name in the answer sentence. Have partners practice describing each shape and the pattern.

Lesson 4 Combine Three-Dimensional Shapes

English Learner Instructional Strategy

Language Structure Support: Modeled Talk

Display the three-dimensional figures: cone, cylinder, cube, and rectangular prism. Have students identify each shape. Review directional words, such as: *on top, above, on, beside, next to, under, underneath,* and *below.* Model each directional word or phrase using the shapes. Distribute geometric solids to each student. Display a cone and a cylinder. Have students identify each shape. Say, *Put a cone on top of a cylinder.* as you model. Have students model and describe using the sentence frame: **The _____ is on top of the _____.** Repeat with other composite shapes.

English Language Development Leveled Activities

Emerging Level	Expanding Level	Bridging Level
Listen and Identify	**Building Oral Language**	**Developing Oral Language**
Distribute 2 to 4 of each solid figure to pairs of students. Direct students to build specific composite shapes. For example, say, *Put a cone on top of a cube,* or *Put two cubes on top of a cylinder.* Allow students time to make the composite shape, then model the correct composite shape for them to compare and self-correct, if necessary. Repeat with new directions until students show understanding.	Distribute 2 of each three-dimensional figure to each student. Have the pairs prop a file folder or book between them to block the view of their partner. One student builds a composite shape and describes how to build the same composite shape to their partner using directional words and the names of the solids. The partition is removed and the two composite shapes are compared. Students exchange roles and repeat.	Distribute newspapers and magazines to groups of four students. Have students search for and cut out images of: cones, cylinders, cubes, and rectangular prisms. Direct students to group the pictures of the figures together to make a variety of composite shapes and glue them onto construction paper. Have each group present their creations to the class using the names of the three-dimensional shapes and directional words.

Multicultural Teacher Tip

Some ELs may be more familiar with an educational system that emphasizes individualized work, learning from lectures, and memorizing facts. As a result, they may be unfamiliar with an approach that stresses collaboration and problem solving. During the Model the Math, be aware of students who are reluctant to communicate or participate. By prompting them with direct questions or by assigning them specific tasks, you can help the students take part in the group dynamic.

NAME _____ DATE _____

Lesson 4 Note Taking
Combine Three-Dimensional Shapes

Read the question. Write words you need help with. Use your lesson to write your Cornell notes. Write or draw math examples to explain your thinking.

Building on the Essential Question	**Notes:**
How can I combine three-dimensional shapes?	I can <u>put</u> <u>together</u> three-dimensional shapes to make composite shapes. I can <u>stack</u> shapes on top of each other.
Words I need help with: See students' words.	It would be difficult to stack a cube on top of a <u>sphere</u>.

My Math Examples:
See students' examples.

Teacher Directions: Read the Building on the Essential Question and have students list words/phrases they need assistance with. Provide descriptions, explanations, or examples of the terms using images or real objects. Read each sentence frame and have students write the appropriate terms. Have students read their notes aloud. Direct students to draw a picture representing the question. Then encourage students to describe their picture to a peer.

Grade 1 · Chapter 10 *Three-Dimensional Shapes* **105**

Dinah Zike Explaining
Visual Kinesthetic Vocabulary®, or VKVs®

What are VKVs and who needs them?

" VKVs are flashcards that animate words by kinesthetically focusing on their structure, use, and meaning. VKVs are beneficial not only to students learning the specialized vocabulary of a content area, but also to students learning the vocabulary of a second language. "

Dinah Zike | Educational Consultant

Dinah-Might Activities, Inc. – San Antonio, Texas

Why did you invent VKVs?

" Twenty years ago, I began designing flashcards that would accomplish the same thing with academic vocabulary and cognates that Foldables® do with general information, concepts, and ideas—make them a visual, kinesthetic, and memorable experience. "

Dinah Zike's
Visual
Kinesthetic
Vocabulary®

I had three goals in mind:

- **Making two-dimensional flashcards three-dimensional**

- **Designing flashcards that allow one or more parts of a word or phrase to be manipulated and changed to form numerous terms based upon a commonality**

- **Using one sheet or strip of paper to make purposefully shaped flashcards that were neither glued nor stapled, but could be folded to the same height, making them easy to stack and store**

Why are VKVs important in today's classroom?

" At the beginning of this century, research and reports indicated the importance of vocabulary to overall academic achievement. This research resulted in a more comprehensive teaching of academic vocabulary and a focus on the use of cognates to help students learn a second language. Teachers know the importance of using a variety of strategies to teach vocabulary to a diverse population of students. VKVs function as one of those strategies. "

Dinah Zike Explaining
Visual Kinesthetic Vocabulary®, or VKVs®

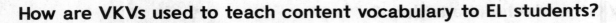

Dinah Zike's
Visual Kinesthetic Vocabulary®

How are VKVs used to teach content vocabulary to EL students?

" VKVs can be used to show the similarities between cognates in Spanish and English. For example, by folding and unfolding specially designed VKVs, students can experience English terms in one color and Spanish in a second color on the same flashcard while noting the similarities in their roots. "

What organization and usage hints would you give teachers using VKVs?

" Cut off the flap of a 6" x 9" envelope and slightly widen the envelope's opening by cutting away a shallow V or half circle on one side only. Glue the non-cut side of the envelope into the front or back of student workbooks or journals. VKVs can be stored in the pocket.

Encourage students to individualize their flashcards by writing notes, sketching diagrams, recording examples, forming plurals (radius: radii or radiuses), and noting when the math terms presented are homophones (sine/sign) or contain root words or combining forms (kilo-, milli-, tri-).

As students make and use the flashcards included in this text, they will learn how to design their own VKVs. Provide time for students to design, create, and share their flashcards with classmates. "

Dinah Zike's book Foldables, Notebook Foldables, & VKVs for Spelling and Vocabulary 4th-12th won a Teachers' Choice Award in 2011 for "instructional value, ease of use, quality, and innovation"; it has become a popular methods resource for teaching and learning vocabulary.

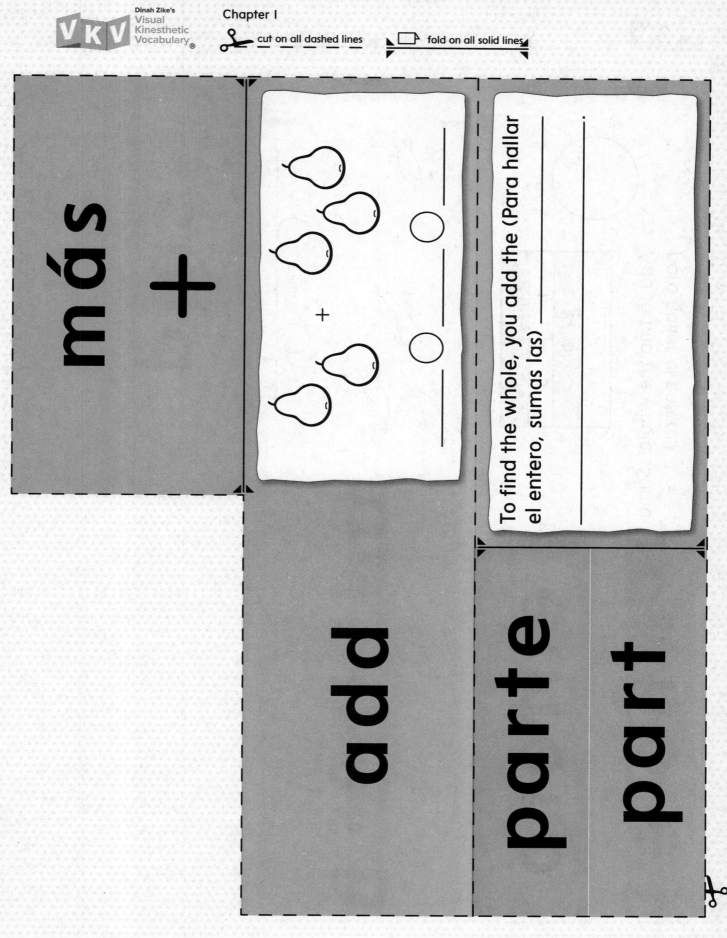

más

+

add

parte

part

To find the whole, you add the (Para hallar el entero, sumas las) _____

Dinah Zike's
Visual
Kinesthetic
Vocabulary®

cut on all dashed lines

fold on all solid lines

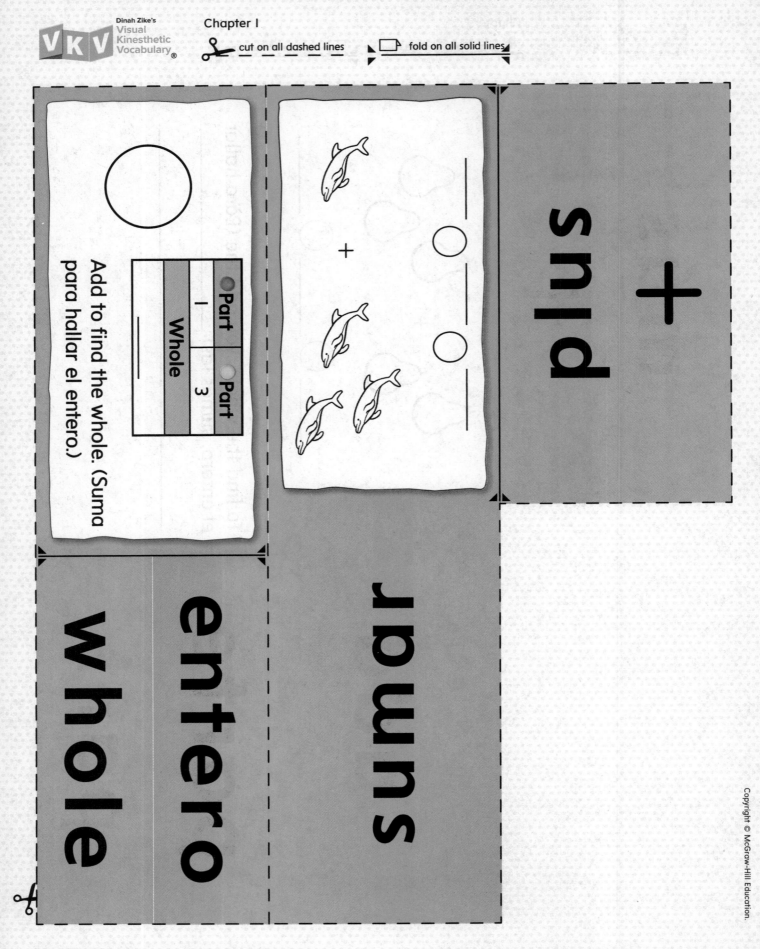

Add to find the whole. (Suma para hallar el entero.)

Part	Whole	Part
I		3

plus

+

whole

entero

sumar

Dinah Zike's
Visual
Kinesthetic
Vocabulary®

Chapter 1

✂ cut on all dashed lines

fold on all solid lines

True or false? (¿Verdadero o falso?)

$2 + 6 = 7$ false (falso)

$4 + 1 = 5$ true (verdadero)   false (falso)

$1 + 8 = 9$ false (falso)

$3 + 4 = 6$ true (verdadero)   false (falso)

true (verdadero)

true (verdadero)

false

zero

true

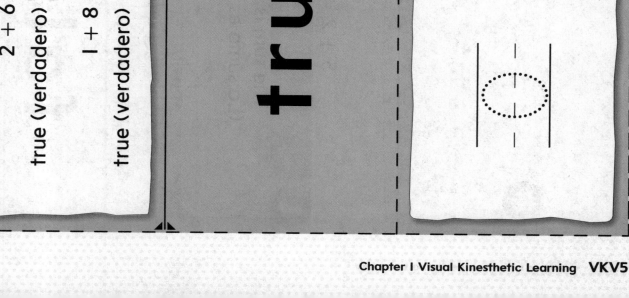

Dinah Zike's
Visual
Kinesthetic
Vocabulary®

Chapter I

✂ cut on all dashed lines

◻ fold on all solid lines

verdadero

falso

Find each sum. (Halla las sumas.)

5 + 0 = ___
0 + 0 = ___
6 + 0 = ___

0 + 3 = ___
9 + 0 = ___
0 + 4 = ___

5 + 2 = 7 ←

The sum is TRUE.
(La suma es VERDADERO.)

5 + 2 = 8 ←

The sum is FALSE.
(La suma es FALSO.)

C

Dinah Zike's
Visual
Kinesthetic
Vocabulary ®

✂ cut on all dashed lines

📄 fold on all solid lines

difference

minus —

Circle the difference. (Encierra
en un círculo la diferencia.)

8 − 5 = 3

Write the subtraction number sentence.
(Escribe el enunciado de resta.)

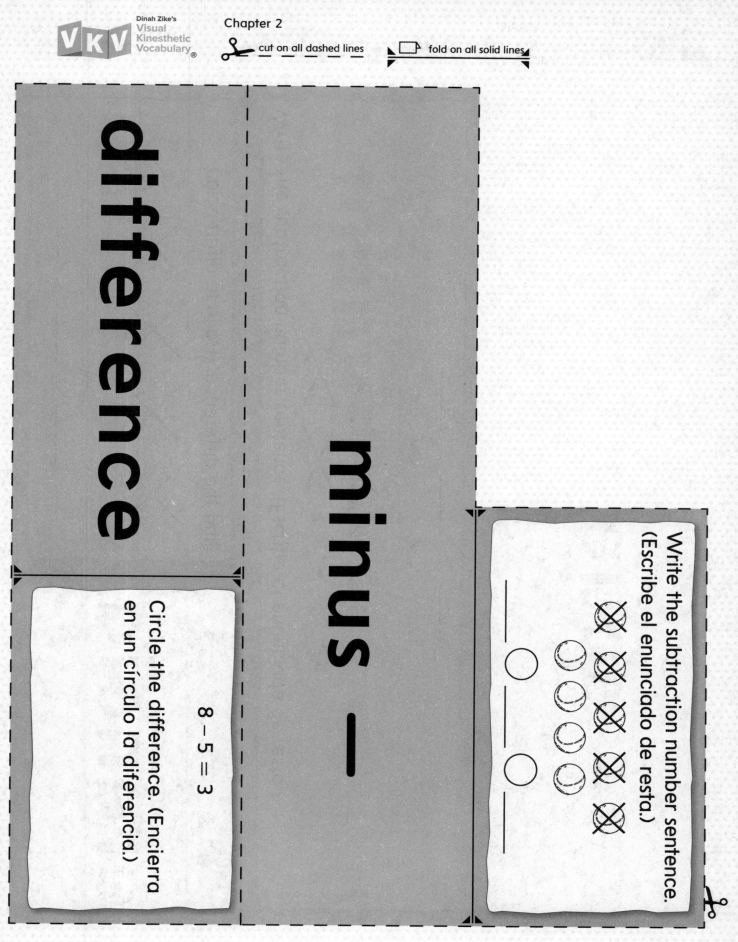

Dinah Zike's
VKV
Visual
Kinesthetic
Vocabulary®

Chapter 2

cut on all dashed lines

fold on all solid lines

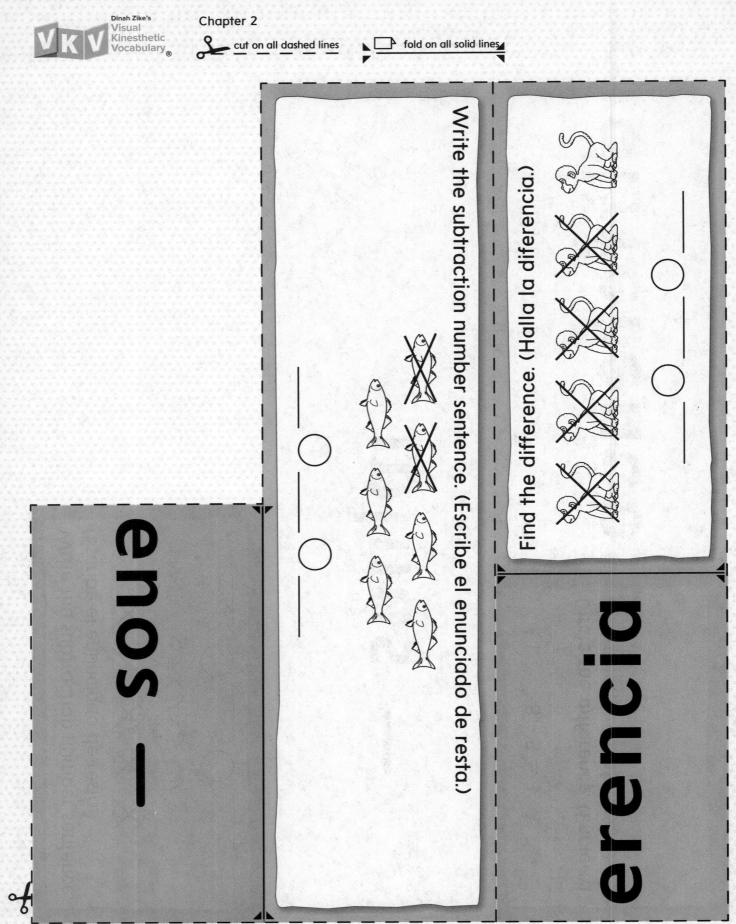

Write the subtraction number sentence. (Escribe el enunciado de resta.)

Find the difference. (Halla la diferencia.)

enos —

erencia

Dinah Zike's
Visual
Kinesthetic
Vocabulary®

V K V

Chapter 2

✂ cut on all dashed lines

▱ fold on all solid lines

related facts

number sentence

Write both facts.
(Escribe ambas operaciones.)

___ + ___ = ___

___ + ___ = ___

Part	Whole		Part
3	5		2

Write a number sentence. (Escribe un enunciado numérico.)

___ ○ ___ ○ ___ = ___

___ + ___ = ___

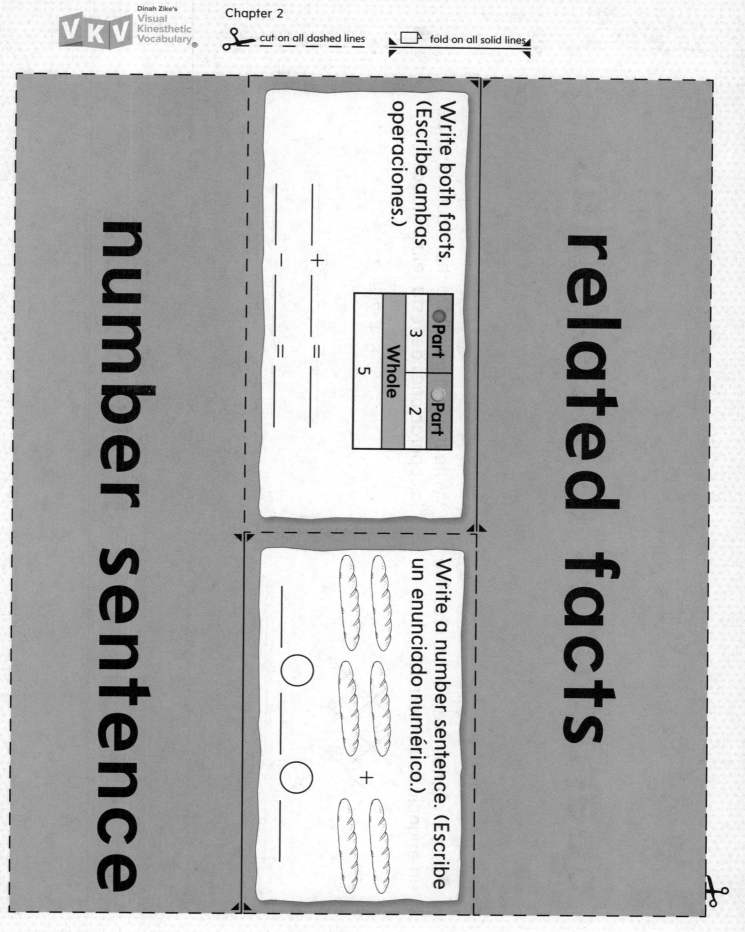

operaciones relacionadas

enunciado numérico

Write the related subtraction facts. (Escribe las operaciones de resta relacionadas.)

$5 + 2 = 7$

___ ___ = ___

___ ___ = ___

Write a number sentence. (Escribe un enunciado numérico.)

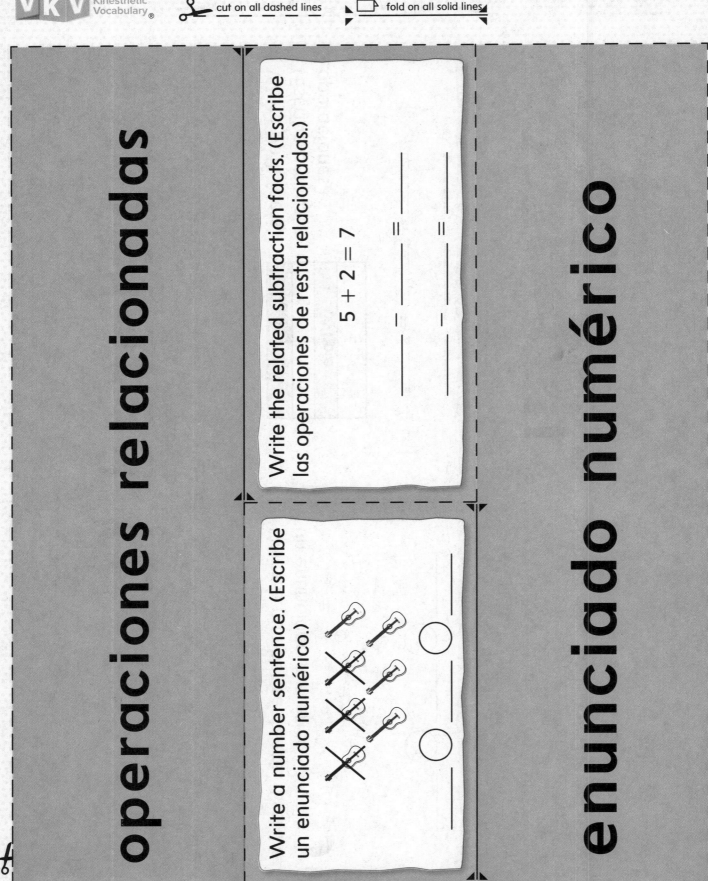

Dinah Zike's
VKV
Visual
Kinesthetic
Vocabulary®

Chapter 3

✂ cut on all dashed lines

▭ fold on all solid lines

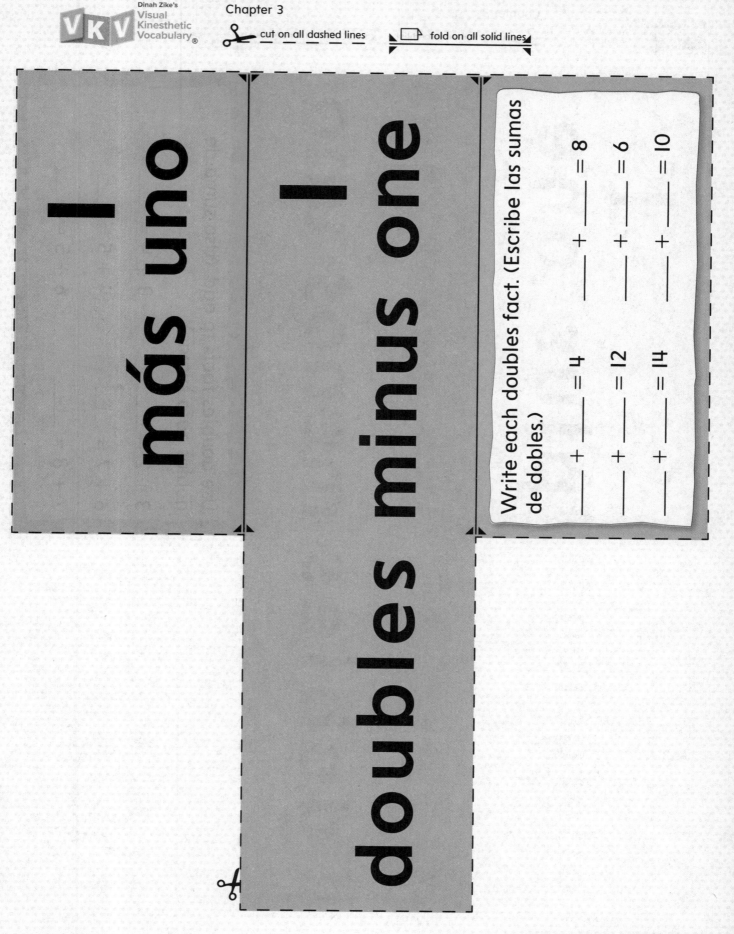

1
más uno

1
doubles minus one

Write each doubles fact. (Escribe las sumas de dobles.)

___ + ___ = 8

___ + ___ = 6

___ + ___ = 10

___ + ___ = 4

___ + ___ = 12

___ + ___ = 14

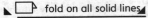 cut on all dashed lines

fold on all solid lines

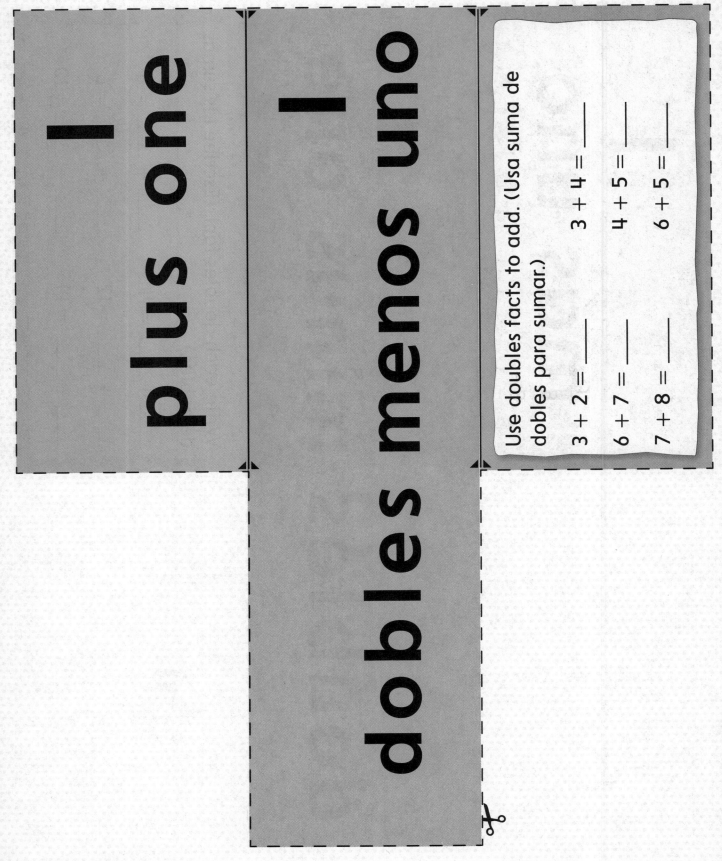

1 plus one

1

dobles menos uno

Use doubles facts to add. (Usa suma de dobles para sumar.)

3 + 2 = ___   3 + 4 = ___

6 + 7 = ___   4 + 5 = ___

7 + 8 = ___   6 + 5 = ___

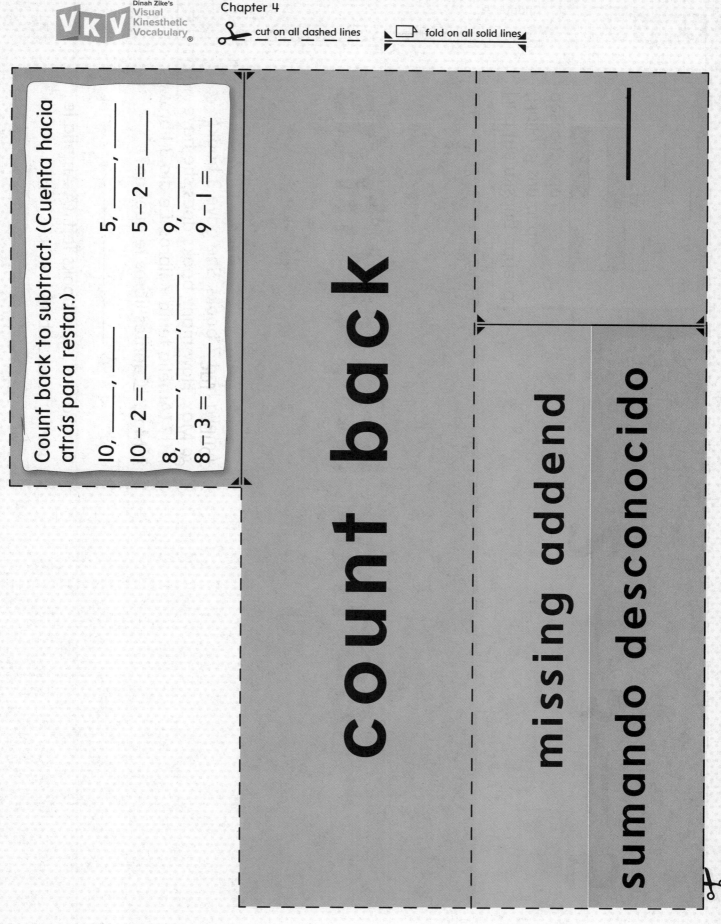

Chapter 4

Count back to subtract. (Cuenta hacia atrás para restar.)

10, _____ , _____   5, _____ , _____

10 – 2 = _____   5 – 2 = _____

8, _____ , _____   9, _____ , _____

8 – 3 = _____   9 – 1 = _____

count back

missing addend

sumando desconocido

Find the missing addend.
(Halla el sumando desconocido.)

● Part | ○ Part

Whole
14

| 6 | 6 |

14 − _____ = 6

_____ + 6 = 14

6 = _____

contar hacia atrás

2 + 3 = 5

Amelia had 7 books. She gave 3 books to Max. How many books does she have left? (Amelia tenía 7 libros. Le dio 3 libros a Max. ¿Cuántos libros le quedan?)

7, _____, _____, _____

Amelia has _____ books left. (A Amelia le quedan _____ libros.)

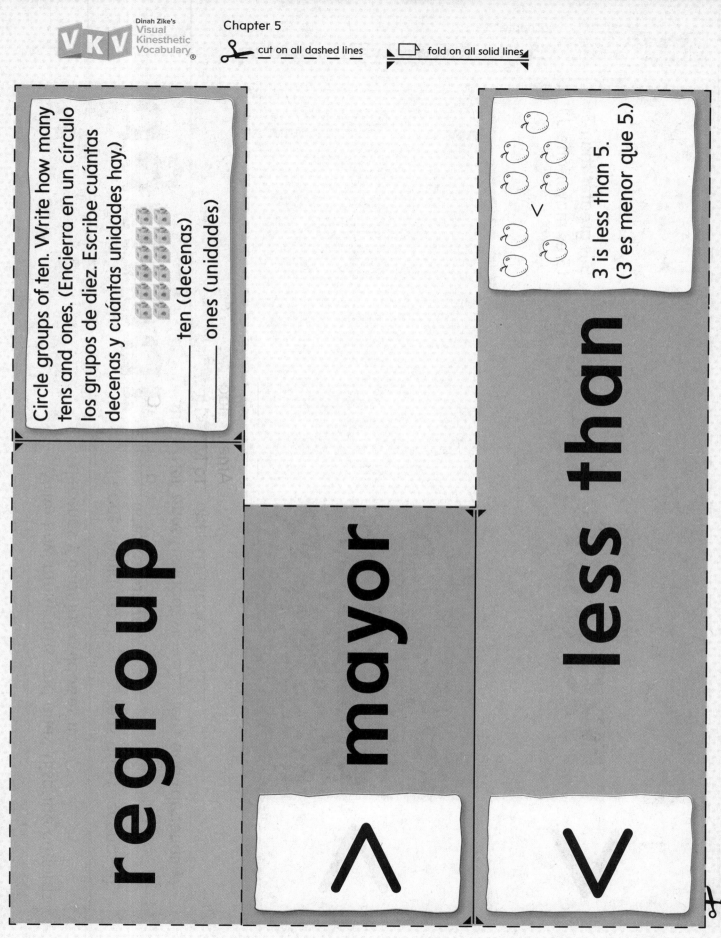

Circle groups of ten. Write how many tens and ones. (Encierra en un círculo los grupos de diez. Escribe cuántas decenas y cuántas unidades hay.)

_____ ten (decenas)

_____ ones (unidades)

3 is less than 5.
(3 es menor que 5.)

regroup

mayor

less than

>

<

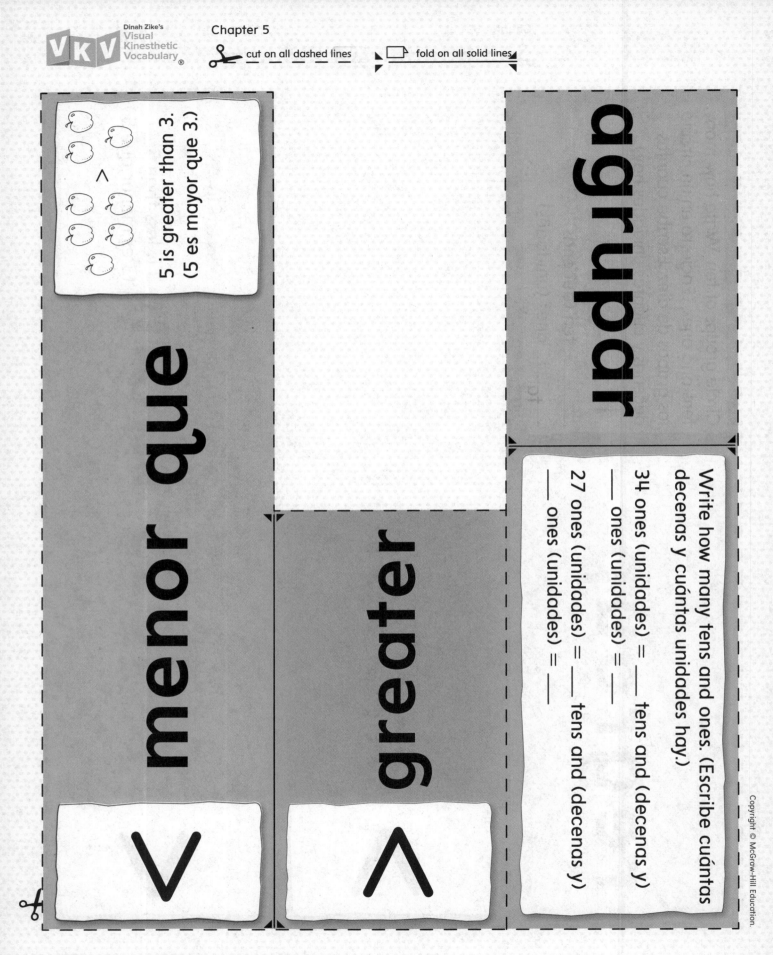

5 is greater than 3.
(5 es mayor que 3.)

agrupar

menor que

greater

Write how many tens and ones. (Escribe cuántas decenas y cuántas unidades hay.)

34 ones (unidades) = ___ tens and (decenas y)
___ ones (unidades) = ___

27 ones (unidades) = ___ tens and (decenas y)
___ ones (unidades) = ___

<

>

Dinah Zike's
Visual
Kinesthetic
Vocabulary®

Chapter 7

✂ cut on all dashed lines

fold on all solid lines

Which is another word for *data*? (¿Cuál es otra palabra para decir *datos*?)

graph (gráfica)

information (información)

tally chart (tabla de conteo)

How many people were surveyed? (¿Cuántas personas fueron encuestadas?)

Favorite Shape		
Shape	Tally	Total
△ Triangle	|||||	
● Circle			
■ Square	|||	

A survey asks people the same (Un encuesta hace a las personas la misma) _____.

The votes of a survey can be marked in a (Los votos de una encuesta pueden marcarse en una) _____.

data

tally chart

survey

cut on all dashed lines
fold on all solid lines

os

What data is shown in the tally chart? (¿Qué datos se muestran en la tabla de conteo?)

Favorite Snack	
Snack	**Tally**
Crackers	||
Bananas	卌 |
Carrots	||||

Write the totals. (Escribe los totales.)

Favorite Pizza Topping		
Topping	**Tally**	**Total**
Cheese	卌 ||	
Pepperoni	|||	
Sausage	||	

tabla de conteo

Write a survey question for the tally chart below. (Escribe una pregunta para una encuesta para la siguiente tabla de conteo.)

Dog						
Cat						
Fish						

encuesta

Dinah Zike's
Visual
Kinesthetic
Vocabulary ®

Chapter 7

✂ cut on all dashed lines

fold on all solid lines

bar graph

con imágenes

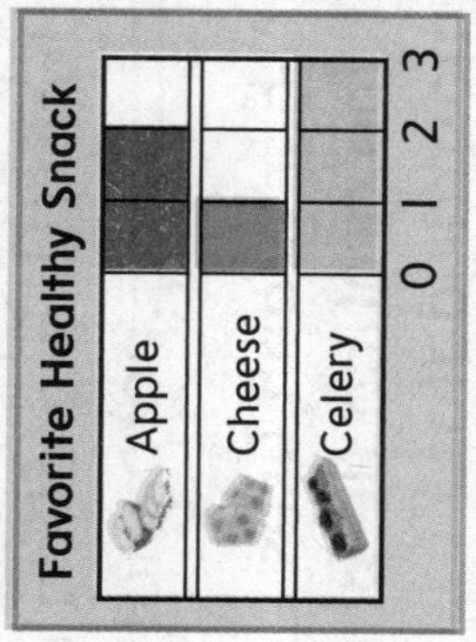

Favorite Healthy Snack

Apple

Cheese

Celery

0 1 2 3

Which snack was the favorite? (¿Cuál merienda era la favorita?)

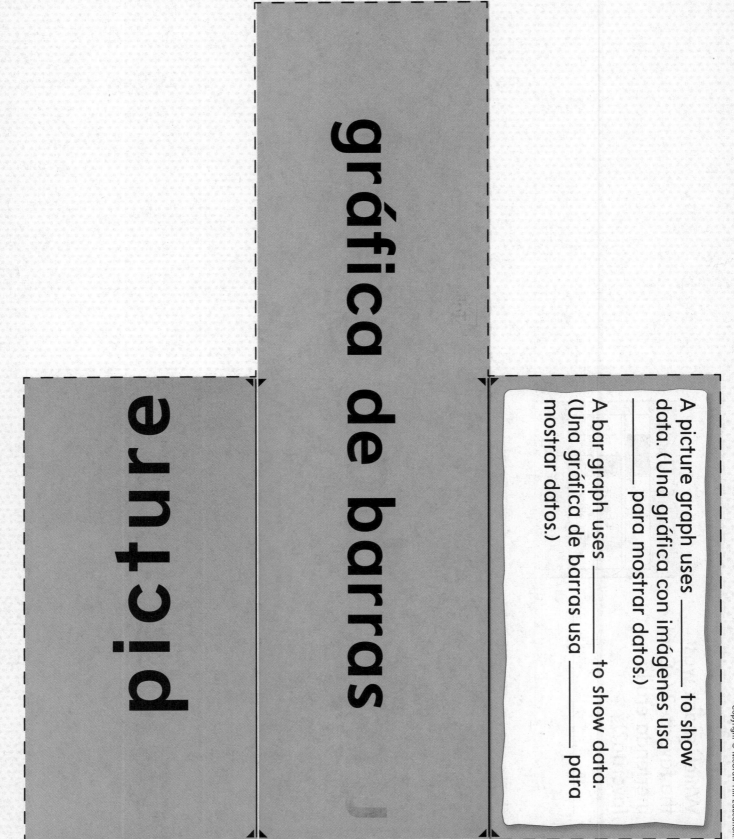

gráfica de barras

picture

gráfica

A picture graph uses —— to show data. (Una gráfica con imágenes usa —— para mostrar datos.)

A bar graph uses —— to show data. (Una gráfica de barras usa —— para mostrar datos.)

Dinah Zike's
Visual
Kinesthetic
Vocabulary®

Chapter 8

✂ cut on all dashed lines

fold on all solid lines

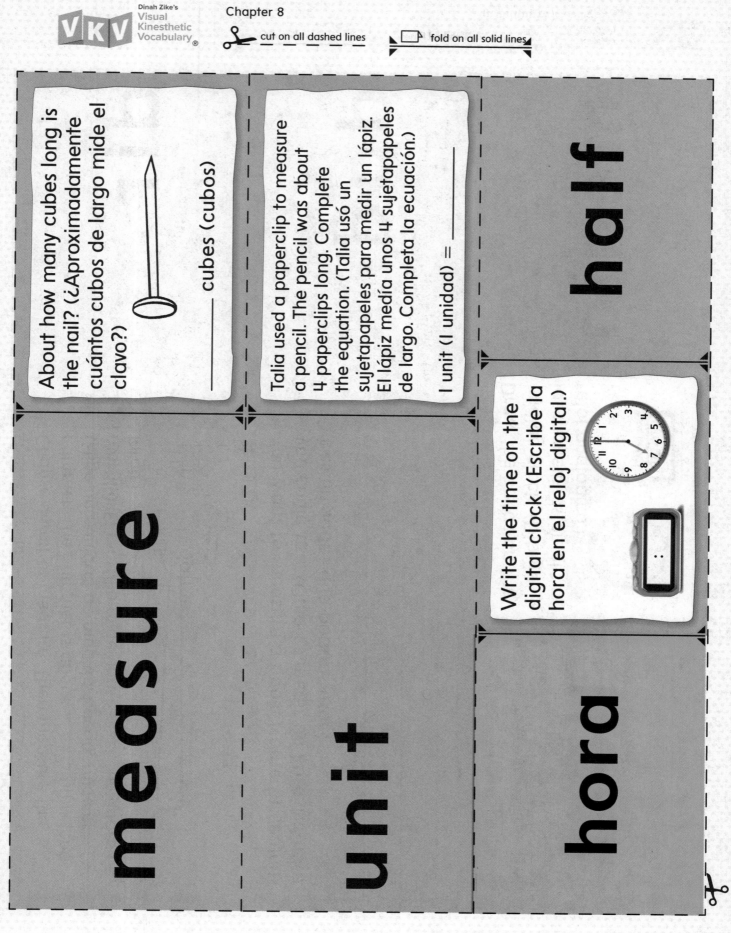

About how many cubes long is the nail? (¿Aproximadamente cuántos cubos de largo mide el clavo?)

_____ cubes (cubos)

Talia used a paperclip to measure a pencil. The pencil was about 4 paperclips long. Complete the equation. (Talia usó un sujetapapeles para medir un lápiz. El lápiz medía unos 4 sujetapapeles de largo. Completa la ecuación.)

1 unit (1 unidad) = _____

Write the time on the digital clock. (Escribe la hora en el reloj digital.)

half

measure

unit

hora

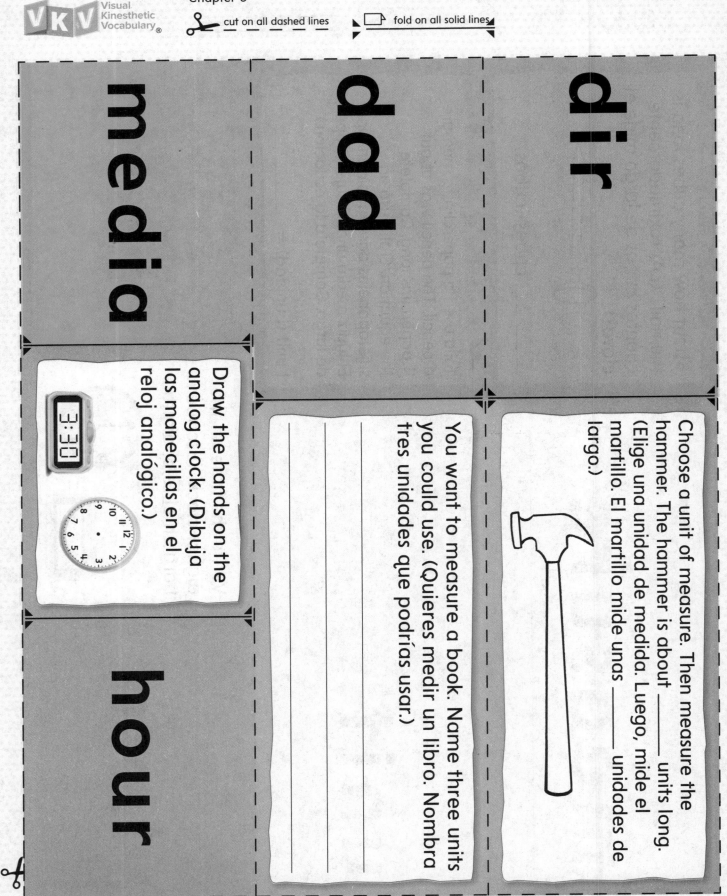

media

dad

dir

hour

Draw the hands on the analog clock. (Dibuja las manecillas en el reloj analógico.)

You want to measure a book. Name three units you could use. (Quieres medir un libro. Nombra tres unidades que podrías usar.)

Choose a unit of measure. Then measure the hammer. The hammer is about —— units long. (Elige una unidad de medida. Luego, mide el martillo. El martillo mide unas —— unidades de largo.)

digital clock

analógico

Malia began reading at 2:00. She read for 1 hour. What time did she stop? Write the time on the digital clock. (Malia empezó a leer a las 2:00. Leyó durante 1 hora. ¿A qué hora terminó de leer? Escribe la hora en el reloj digital.)

reloj digital

analog

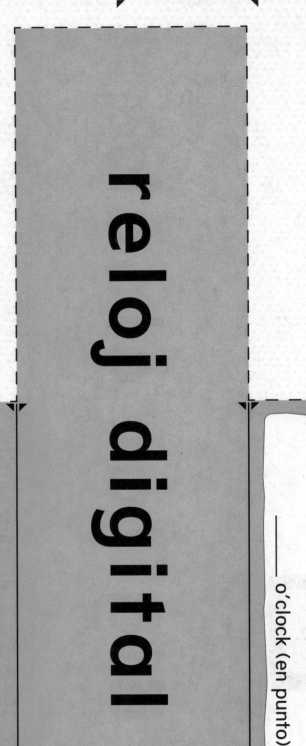

What time is shown on the analog clock?
(¿Qué hora muestra el reloj analógico?)

_____ o'clock (en punto)

Dinah Zike's Visual Kinesthetic Vocabulary ®

minute hand

horaria

minute hand

Draw the minute hand to show 11:00. (Dibuja la manecilla de los minutos para mostrar las 11:00.)

hour

manecilla de los minutos

Draw the hour hand to show 4:00. (Dibuja la manecilla horaria para mostrar las 4:00.)

Dinah Zike's
Visual
Kinesthetic
Vocabulary®

Chapter 9

✂ cut on all dashed lines   fold on all solid lines

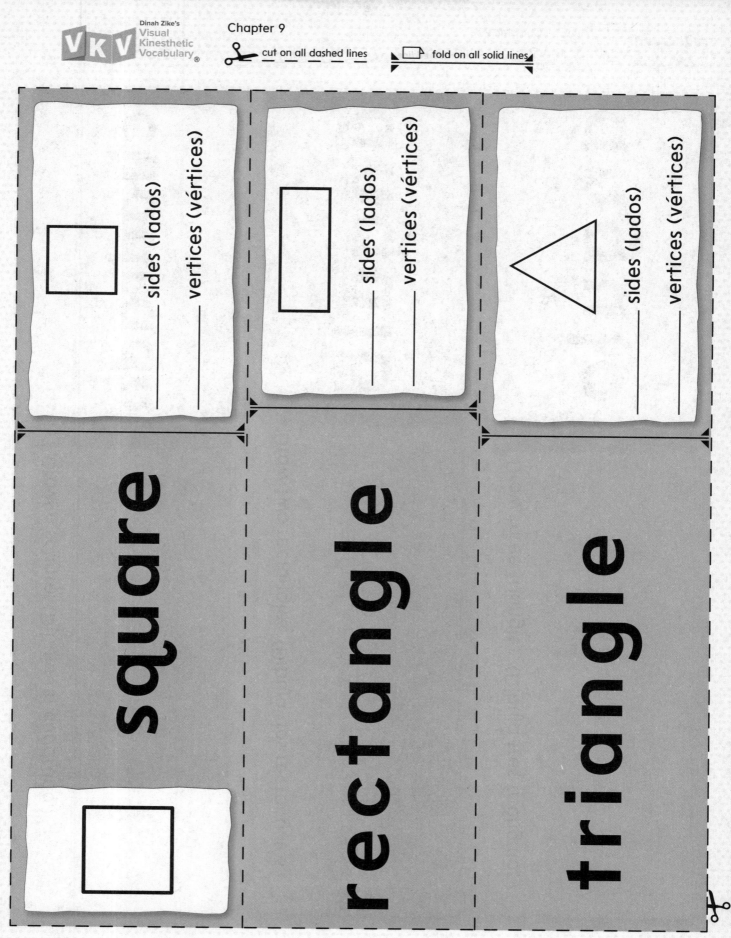

sides (lados) ____

vertices (vértices) ____

sides (lados) ____

vertices (vértices) ____

sides (lados) ____

vertices (vértices) ____

square

rectangle

triangle

Dinah Zike's
Visual
Kinesthetic
Vocabulary®

✂ cut on all dashed lines

▷ fold on all solid lines

ángulo

ángulo

cuadrado

Draw three triangles. (Dibuja tres triángulos.)

Draw two rectangles. (Dibuja dos rectángulos.)

Draw a square. (Dibuja un cuadrado.)

Dinah Zike's
VKV **Visual
Kinesthetic
Vocabulary**®

Chapter 9

✂ cut on all dashed lines   ⬚ fold on all solid lines

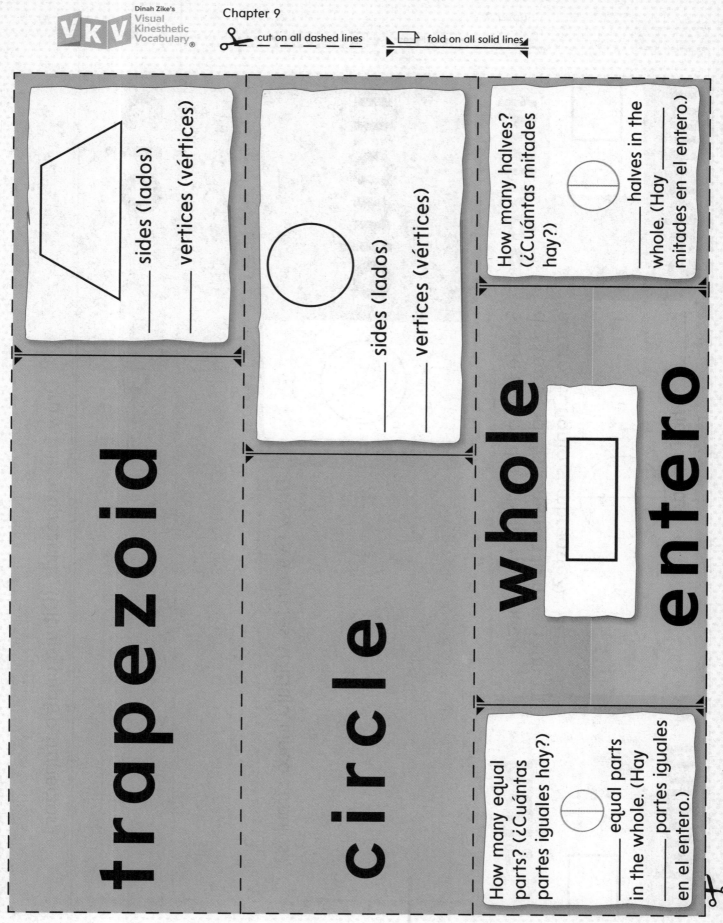

sides (lados)

vertices (vértices)

sides (lados)

vertices (vértices)

How many halves? (¿Cuántas mitades hay?)

____ halves in the whole. (Hay ____ mitades en el entero.)

How many equal parts? (¿Cuántas partes iguales hay?)

____ equal parts in the whole. (Hay ____ partes iguales en el entero.)

trapezoid

circle

whole

entero

Chapter 9

✂ cut on all dashed lines   ⬜ fold on all solid lines

Dinah Zike's
Visual
Kinesthetic
Vocabulary®
VKV

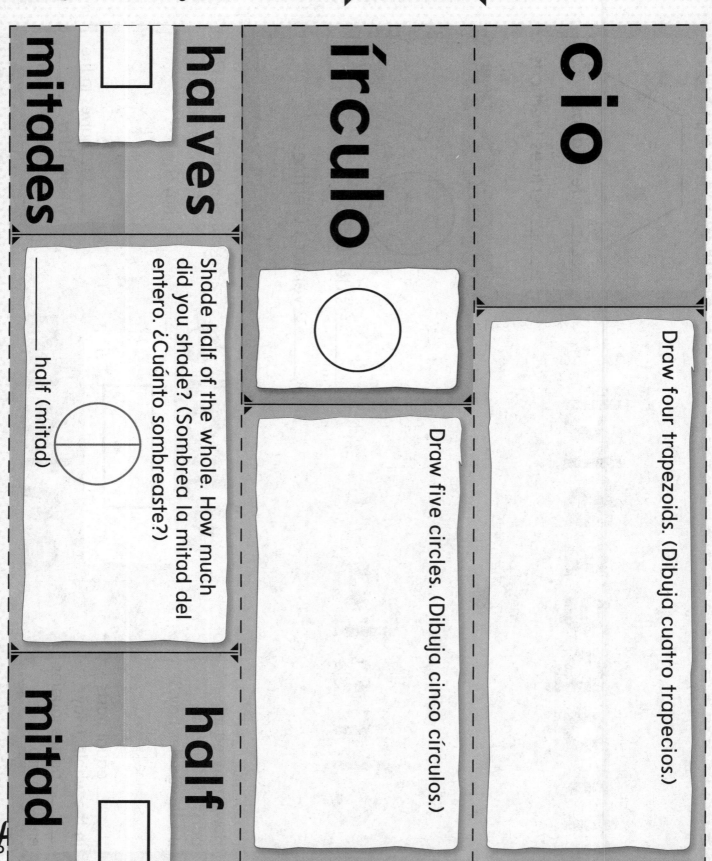

mitades

halves

írculo

cio

Shade half of the whole. How much did you shade? (Sombrea la mitad del entero. ¿Cuánto sombreaste?)

—— half (mitad)

Draw five circles. (Dibuja cinco círculos.)

Draw four trapezoids. (Dibuja cuatro trapecios.)

mitad

half

✂ cut on all dashed lines

fold on all solid lines

three-dimensional

rectangular

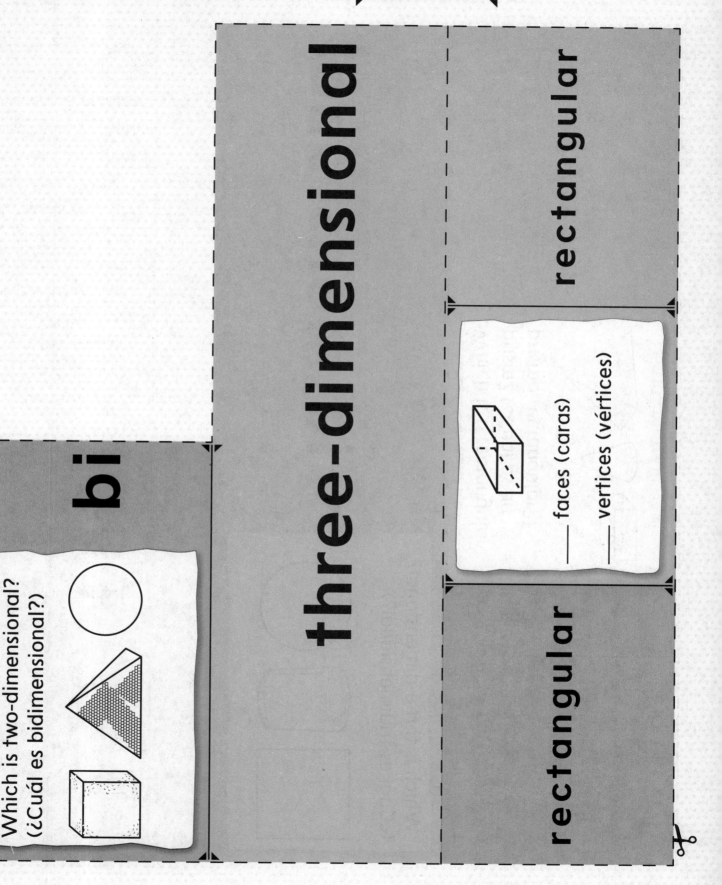

—— faces (caras)

—— vertices (vértices)

rectangular

bi

Which is two-dimensional?
(¿Cuál es bidimensional?)

Dinah Zike's
VKV Visual
Kinesthetic
Vocabulary®

Chapter 10

✂ cut on all dashed lines   ⬛ fold on all solid lines

prisma

tridimensional

prism

two-

Which is a rectangular prism? (¿Cuál es un prisma rectangular?)

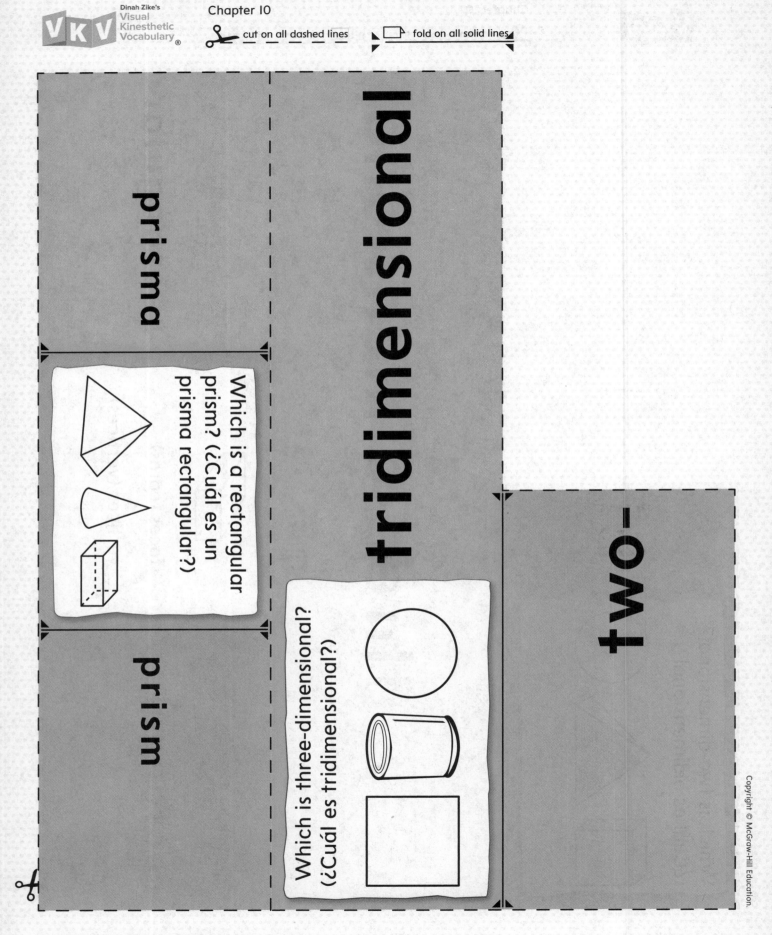

Which is three-dimensional? (¿Cuál es tridimensional?)

VKV Dinah Zike's **Visual Kinesthetic Vocabulary** ®

✂ cut on all dashed lines

✂ fold on all solid lines

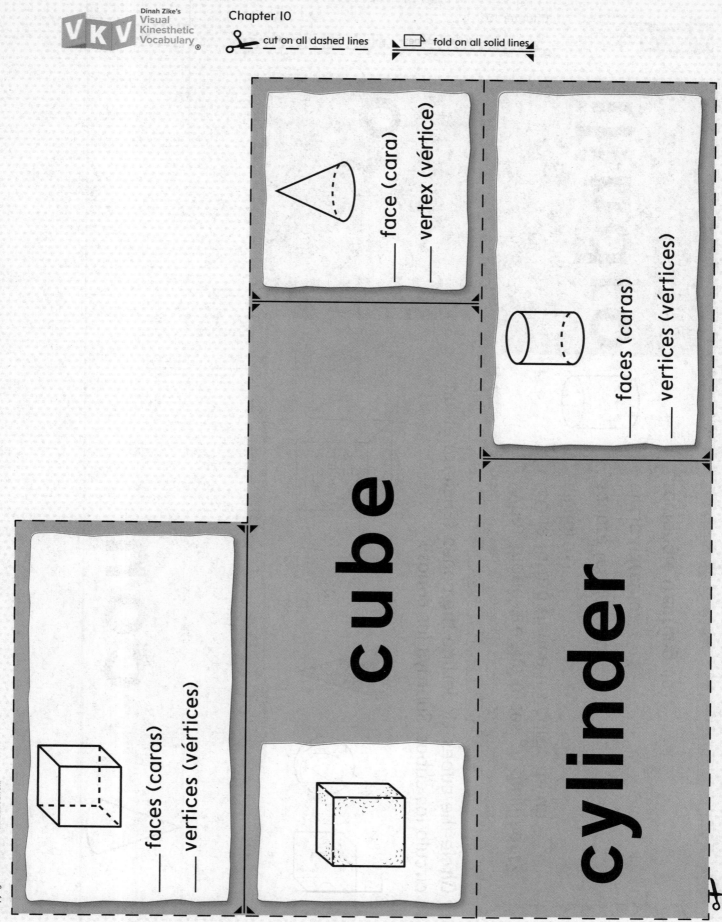

face (cara)

vertex (vértice)

faces (caras)

vertices (vértices)

cube

faces (caras)

vertices (vértices)

cylinder

ilindro

o

con

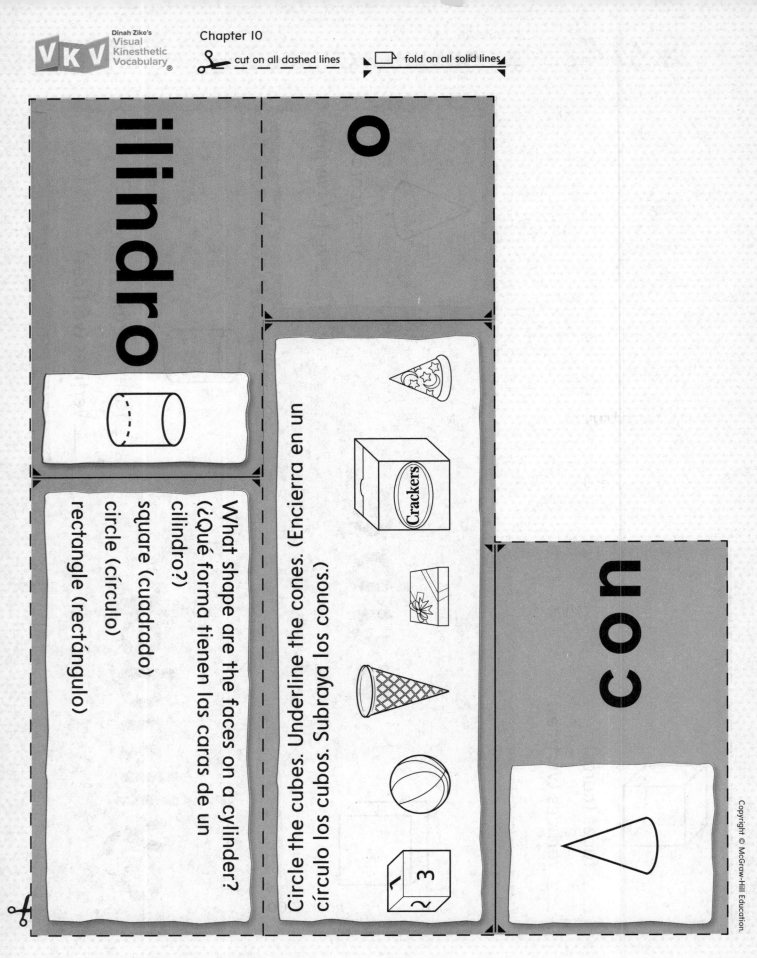

What shape are the faces on a cylinder? (¿Qué forma tienen las caras de un cilindro?)

square (cuadrado)

circle (círculo)

rectangle (rectángulo)

Circle the cubes. Underline the cones. (Encierra en un círculo los cubos. Subraya los conos.)

Crackers

3 2 1

VKV Answer Appendix

Chapter 1
VKV3
part: parts
addition: $2 + 3 = 5$

VKV4
whole: 4
sumar: $3 + 1 = 4$

VKV5
true/false: false; true; true; false
zero: See students' work.

VKV6
cero: 5; 3; 0; 9; 6; 4

Chapter 2
VKV7
minus: $9 - 5 = 4$
difference: 3

VKV8
menos: $7 - 2 = 5$
diferencia: $5 - 4 = 1$

VKV9
number sentence: $4 + 2 = 6$
related facts: $3 + 2 = 5$; $5 - 3 = 2$
or $5 - 2 = 3$

VKV10
enunciado numérico: $7 - 3 = 4$
operaciones relacionadas: $7 - 2 = 5$;
$7 - 5 = 2$

Chapter 3
VKV11
doubles minus one: $2 + 2$; $4 + 4$; $6 + 6$;
$3 + 3$; $7 + 7$; $5 + 5$

VKV12
dobles minos uno: 5; 7; 13; 9; 15; 11

Chapter 4
VKV13
count back: 9, 8, 8; 4, 3, 3; 7, 6, 5, 5;
8, 8
missing addend: $2 + 3 = 5$: 8; 8

VKV14
contar hacia atrás: 6, 5, 4, 4 books left

Chapter 5
VKV15
regroup: 1 ten, 2 ones

VKV16
reagrupar: 3 tens and 4 ones = 34;
2 tens and 7 ones = 27; 5 tens and
3 ones = 53

Chapter 7
VKV17
data: information
tally chart: 9
survey: question; tally chart

VKV18
datos: votes for favorite kind of snack
tabla de confeo: 7; 3; 2
encuesta: Sample answer: What is your
favorite pet?

VKV19
bar graph: celery

VKV20
gráfico de barras: pictures; bars

Chapter 8
VKV21
measure: See students' work.
unit: 1 paperclip
hour: 7:00

VKV22

medir: See students' work.
half: See students' work.

VKV23

digital clock: 3:00

VKV24

reloj digital: 5 o'clock

VKV25

minute hand: See students' work.

VKV26

manecilla de los minutos: See students' work.

Chapter 9

VKV27

square: 4 sides, 4 vertices
rectangle: 4 sides, 4 vertices
triangle: 3 sides, 3 vertices

VKV28

cuadrado: See students' work.
rectángulo: See students' work.
triángulo: See students' work.

VKV29

trapezoid: 4 sides, 4 vertices
circle: 0 sides, 0 vertices
whole: 2 equal parts; 2 halves

VKV30

trapecio: See students' work.
círculo: See the students' work.
halves/half: 1 half

Chapter 10

VKV31

three-dimensional: See students' work.
rectangular prism: 6 faces, 8 vertices

VKV32

tridimensional: See students' work.
prisma rectangular: See students' work.

VKV33

cube: 1 face, 1 vertex; 6 faces, 8 vertices
cylinder: 2 faces, 0 vertices

VKV34

cono: See students' work.
cilindro: circle